Ons Haddad
Yosr Kadri
Yassmine Maatouk

The bacterial world and COVID-19

Ons Haddad
Yosr Kadri
Yassmine Maatouk

The bacterial world and COVID-19

What are the special features of intensive care?

ScienciaScripts

Cover image: www.ingimage.com

This book is a translation from the original published under ISBN 978-620-6-72579-4.

Publisher:
Sciencia Scripts
is a trademark of
Dodo Books Indian Ocean Ltd. and OmniScriptum S.R.L publishing group

120 High Road, East Finchley, London, N2 9ED, United Kingdom
Str. Armeneasca 28/1, office 1, Chisinau MD-2012, Republic of Moldova, Europe
Printed at: see last page
ISBN: 978-620-8-31841-3

Contents

INTRODUCTION

Bacterial co-pathogens are frequently identified in viral respiratory tract infections such as influenza, and represent a major cause of morbidity and mortality, requiring rapid diagnosis and antibacterial treatment.

Although highly variable, bacterial co-infection in patients with severe influenza can be as high as 20-30% **[1,2]** and is associated with increased severity of illness and increased risk of death. The prevalence, incidence and characteristics of bacterial infection in patients infected with severe acute respiratory syndrome coronavirus 2 (SARS-CoV-2) are not well studied and are considered a major knowledge gap.

Although antibiotics are ineffective in the treatment of coronoavirus 2019 (COVID-19), they are prescribed for patients with COVID-19 for a variety of reasons. In particular, it is difficult to rule out a bacterial co-infection. Based on concerns about increased mortality in patients with bacterial superinfection during influenza pandemics, several guidelines recommend the use of empirical antibiotics for patients with severe COVID-19 **[3,4]**. However, this assumption raises concerns about antibiotic overuse and the subsequent harm associated with bacterial resistance.

Understanding the likelihood of acute respiratory bacterial co-infection[1] in patients with COVID-19, and the causative pathogens, is therefore essential for treating patients with COVID-19 and for contributing to the reasonable use of antibiotics and minimising the adverse consequences of overuse. Indeed, this knowledge could have a significant impact on recommendations for the management of empirical antibiotic therapy for patients with COVID-19.

In this context, the aims of this study were to describe the epidemiological and bacteriological characteristics of positive microbiological samples taken from patients with COVID-19, in relation to the pre-COVID period, in these people hospitalised in the polyvalent intensive care unit at the Fattouma Bourguiba Monastir University Hospital.

CHAPTER 1

1. BIBLIOGRAPHIC REMINDER

1.1. Presentation of SARS-CoV2

The Covid-19 pandemic is a highly contagious respiratory disease **[5,6]**. From its declaration in December 2019 until 5 July 2023, it has infected 767,726,861 people, resulting in 6,948,764 deaths worldwide **[7]**.

The usual symptoms of COVID-19 mimic those of influenza and include fever, cough, shortness of breath, dry cough, malaise and, in some cases, changes in taste and smell **[8]**. The severity of this disease can range from asymptomatic illness to acute respiratory distress syndrome[1] (ARDS) and death **[9]**.

The coronavirus (CoV) was first identified in 1966 by Tyrell and Bynoe **[10]**. CoV **(Figure 1)** can affect both animals and humans. The structure of the virus is known for its typical spherical morphology with the core of the envelope and projections of glycoproteins from their envelope, giving them the appearance of a "crown", hence the name coronavirus **[11]**. Nucleocapsids have helical symmetry, with a positive RNA genome located in the envelope, giving them the largest genome of all RNA viruses. SARS-CoV-2 belongs to the order Nidovirales, family Coronaviridae, subfamily Orthocoronavirinae, genus Betacoronavirinae and subgenus Sarbecovirus. Genetic recombination within members of the same or different genera leads to increased genetic diversity, which results in spread to other species and the emergence of new viruses **[12]**.

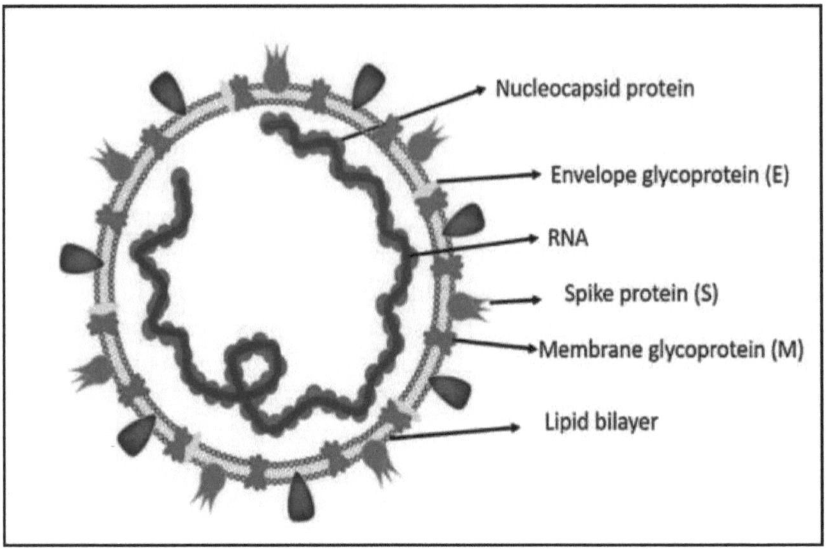

Figure 1: Structure of the human coronavirus responsible for respiratory syndrome [13].

It is important to determine the source of origin and transmission in order to develop preventive strategies to contain the infection. In the case of SARS-CoV, researchers initially focused on raccoon dogs and palm civets as the key reservoir of infection. However, only samples isolated from civets on the food market gave positive results for the detection of viral RNA, suggesting that palm civets could be secondary hosts **[13]**.

According to current data, the COVID-19 virus is transmitted mainly between individuals via respiratory droplets and contact routes **[14]**.

1.2. Co-infections and microbial superinfection in COVID-19

Co-infection refers to the simultaneous infection of a cellular host by two or more pathogen species and/or strains, whereas superinfection is a scenario in which one pathogen infects the host some time before infection by the second pathogen **[15]**.

In both cases, the fate of the infected host often depends on a balance between the host's protective immunity and immunopathology **[16]**. The ubiquity or incidence of co-infection in humans is unknown, but it is sometimes thought to be more common than isolated infection **[17]**. Co-infecting pathogens may alter the population of the primary pathogen. For example, Van der Hoek et al. reported that in respiratory co-infections, the load of human coronavirus (hCoV) was much lower than in a single infection **[18]**.

Co-infections can be caused by multiple infectious agents of viral, bacterial, archaeal and fungal origin **(Figure 2)**, and appear to occur simultaneously with the initial onset of disease **[19]**.

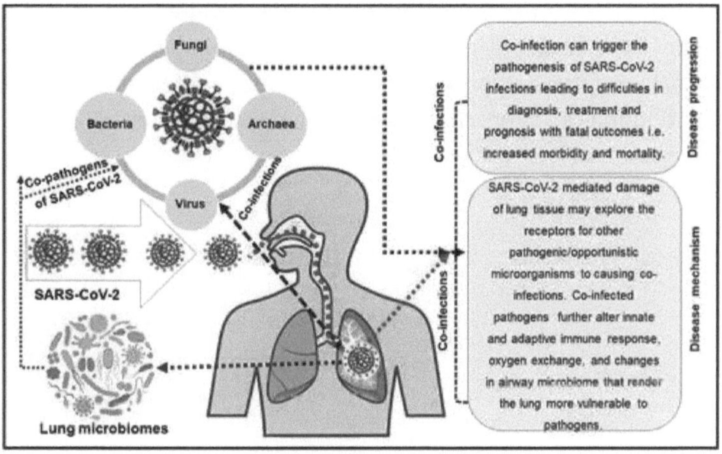

SARS-CoV-2: severe acute respiratory syndrome coronavirus 2

Figure 2: Interrelationship between SARS-CoV-2 and respiratory microbiomes leading to co-infections in patients COVID-19 [20].

COVID-19 patients may be co-infected with different microbial domains, including viruses (other than SARS-CoV-2), bacteria, archaea and fungi. These diverse microbial communities simultaneously complicate the pathophysiology and course of SARS-CoV-2 infections.

The morbidity of co-infections has already been studied in some cohorts, and is often reported to be worse than that of single infections. Recently, several observational and cohort studies have reported that pulmonary complications occurred in 51.2% of patients with COVID-19, including 82.6% death, and that the independent risk factors for mortality were male sex and Page aged 65 years or older **[21]**. However, the occurrence of co-infection in deaths across age and sex cohorts of COVID-19 patients has not yet been studied. Co-infections in COVID-19 may not be captured by age group, and unexplained heterogeneity in the severity of COVID-19 may be due to differences between studies in disease severity, patient comorbidities, differences in treatment (such as corticosteroid administration), antibiotic use before and during hospitalisation, or other unidentified covariates. However, underestimation of secondary bacterial or fungal infections probably develops later in the course of COVID-19 [20].

1.3. Possible pathophysiology of secondary bacterial infection

The epithelium of the airways **(Figure 3)** is generally protected by a layer of mucus. Mucins, types of charged glycoproteins, are the main components of mucus. The most common types of mucin in human sputum are MUC5AC and MUC5B, and they support the innate immune system with their antiviral and anti-inflammatory properties, trapping and clearing viruses **[22]**.

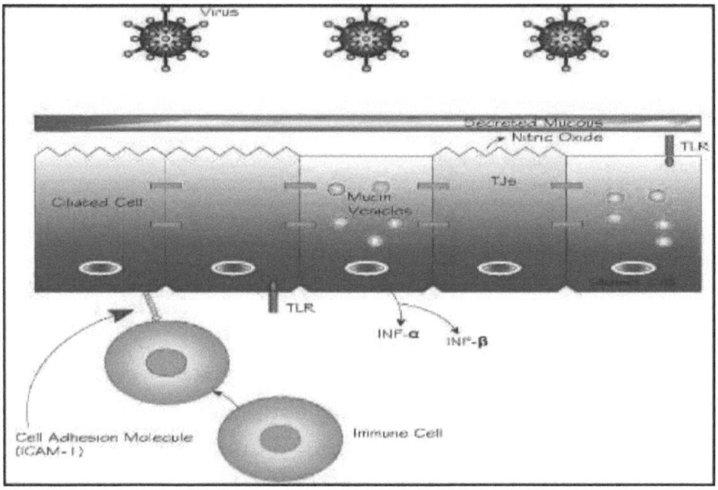

TLR: Toll-Like Receptors; INF: Interferon; ICAM-I: Intercellular Adhesion Molecule-1.

Figure 3: Modes of protection of the airway epithelium [23].

On the other hand, overproduction of these mucins could have a counterproductive effect **[24]**. In addition to its physical protection role, the airway epithelium also recognises micro-organisms via pattern recognition receptors such as RIG (retinoic acidinducible gene) type helicases, TLR (toll-like receptors) type receptors and NLR (nucleotidebinding oligomerization domain) type receptors **[23]**. NLRs and RIG-type helicases trigger innate immune responses by cytosolic detection of bacterial and viral components **[25]**. Nod1 and Nod2, which are members of the NLR family, are triggered by molecules synthesised during the assembly and/or degradation of bacterial peptidoglycan **[26]**.

1.4. Effect of the virus on the immune system

The ongoing post-viral desensitisation of lung cells to TLR signals may contribute to the secondary bacterial pneumonia associated with viral infection. For example, the TLR4 and TLR5 pathways are deformed following viral infection, leading to a weakening of the neutrophil attraction mechanism, and therefore increased attachment of *Streptococcus pneumonia* and *Pseudomonas aeruginosa* to airway epithelial cells **[25]**. The interaction between a virus and the host cell during a disease induces immune responses that include the generation of pro-inflammatory molecules. Despite their important role as bactericides, pro-inflammatory cytokines such as TNF-a synthesised in response to disease may be harmful to host cells **[27]**. During viral disease, activation of TLR and RIG-I-like receptors results in the synthesis of type I IFNs, which can enhance the inflammatory response to TLR ligands, including lipopolysaccharide (LPS) **[28]**.

In addition, certain bacteria such as *Staphylococcus aureus* incorporate into respiratory epithelial cells (human alveolar basal cells with adenocarcinoma) during respiratory viral disease by stimulating ICAM-1 expression **[29]**.

Finally, the interaction between Nod1/Nod2 and type I IFNs leads to recognition of bacteria, but indicates harmful effects on host cells **[30]**.

Table I shows the bacterial co-pathogens commonly reported in the case of COVID-19, their mode of transmission, the possible mechanism of co-infections and the results.

Table I: Microbial co-pathogens frequently reported in the context of COVID-19, their mode of transmission, possible mechanism of co-infection and results [20].

Type of co-infection	Co-pathogenes	Transmission route	Person-to-person transmission	Possible mechanisms of co-infection and pathogenesis	Possible results
Bacterian	*Streptococcus pneumoniae*	Respiratory	Yes	Normal opportunistic flora of the human upper respiratory tract	Severe respiratory distress followed by pleural effusion and necrotizing pneumonia, with a higher mortality rate.
	Staphylococcus aureus	Respiratory/digestive/contact	Yes	Normal opportunistic flora of the human upper respiratory tract, intestinal mucosa and skin	Pneumonic necrosante. Bacteria and higher mortality.
	Pseudomonas aeruginosa	Contact	Yes	Opportunistic pathogen causing HCAI, mainly associated with poor hygiene, mechanical ventilation and	ND

				urinary catheterisation.	
	Acinetobacter baumannii	Contact	Yes	Mechanical ventilation	ND
	Klebsiella pneumoniae	Respiratory/Contact	Yes	Normal opportunistic flora of the human mouth, skin and intestines	Deadly septicemic.
	Mycoplasma pneumoniae	Respiratory/contact	Yes	ND	Pneumonic severe. Increased morbidity, mortality and severity of disease.
	Clamydia pneumoniae	Respiratory/contact	Yes	ND	Pneumonic severe.

HCAI: Healthcare Associated Infections; COVID-19: Coronavirus 2019; SARS-CoV-2: Severe Acute Respiratory Syndrome Coronavirus 2; ND: Non ddtermind

Table I: Microbial co-pathogens frequently reported in the context of COVID-19, their mode of transmission, possible mechanism of co-infection and results [20].

Type of co-infection	Co-pathogenes	Transmission route	Person-to-person transmission	Possible mechanisms of co-infection and pathogenesis	Possible results
	Legionella pneumophila	Digestive/Respiratory	Yes	ND	Elevation of aspartate aminotransferase, blood urea nitrogen, creatinine, lactate dehydrogenase and C-reactive protein.
	Haemophilus influenzae	Respiratory/contact	Yes	Normal opportunistic flora of the human upper respiratory tract	ND
	Neisseria meningitides	Respiratory/contact	Yes	ND	Convulsion, elevated C-reactive protein, headache, stiff neck, chills, confusion and new purpuric eruption on hands and feet.
	Mycobacterium tuberculosis	Respiratory	Yes	The cytokine storm produced by COVID-19 can reactivate latent tuberculosis or stimulate the development of active tuberculosis. Pulmonary lesions	Co-infection is associated with both severity and rate of disease progression. A 2.2-fold higher risk of death and a 25% lower risk of recovery were reported. In addition, a shorter time to death

				caused by tuberculosis may also increase the severity of disease caused by SARS-CoV-2.	and a longer time to recovery were observed.

HCAI: Healthcare Associated Infections; COVID-19: Coronavirus 2019; SARS-CoV-2: Severe Acute Respiratory Syndrome Coronavirus 2; ND: Non ddtermind

1.5. Treatment

Prevention is the best way to avoid infection with COVID-19. The World Health Organisation recommends a number of hygiene and personal practices, such as regular hand washing with soap and water, the use of hand rubs based on ethyl alcohol (ethanol), isopropyl alcohol (isopropanol, propan-2-ol) and n-propanol, respiratory hygiene and social distancing **[31]**.

SARS-COV2 has a high frequency of mutation and genomic recombination, which represents a major challenge for the synthesis of an effective vaccine **[32]**.

Antimicrobial therapy could play a crucial role in the treatment of confirmed or suspected respiratory fungal or bacterial co-infection. This therapy may be empirical or targeted in infected patients. The respiratory symptoms of COVID-19 and chest radiographic changes can mimic those of atypical bacterial pneumonia and can therefore make it very difficult for clinicians to interpret **[33]**. Therefore, empiric antimicrobial therapy is appropriate for bacterial/fungal pneumonia in these patients **[23]**.

Antibiotic therapy has shown positive therapeutic results in various clinical studies **[34]**. Cephalosporins, carbapenems and quinolones are good initial choices for patients with COVID-19 in conjunction with antiviral agents **[35]**. Since empirical antibiotics may be beneficial, it is crucial to bear in mind the side effects, and the spread of antimicrobial resistance **[23]**. Clinicians concerned about secondary infections should refer to institutional susceptibility testing and/or take into account local drug resistance and hospital epidemiological patterns when choosing empirical regimens **[23]**. One possible solution to support antimicrobial prescribing in COVID-19 is the use of bacterial-specific biomarkers, such as procalcitonin **[36]**. Indeed, procalcitonin has been shown to help differentiate between bacterial and viral infections **[36]**.

CHAPTER 2

2. MATERIALS AND METHODS

2.1. Equipment

This is a retrospective study carried out in the intensive care unit at the Fattouma Bourguiba University Hospital in Monastir over a 2-year period from 1er April 2020 to 31 March 2022.

2.1.1. Inclusion criteria

Patients aged 18 or over hospitalised in the general intensive care unit for acute respiratory failure[1] secondary to infection with SARS-CoV2 were included. The diagnosis of this infection was confirmed by a positive RT-PCR for SARS-CoV2 on a nasopharyngeal sample or a sample from the lower respiratory tract if the patient was on mechanical ventilation.

2.1.2. Exclusion criteria

All patients hospitalised for less than 48 hours in the general intensive care unit at the Fattouma Bourguiba University Hospital in Monastir were excluded.

2.1.3. Data collection

Patient data were collected using data collection forms filled in from the medical records of patients in the intensive care unit, including for each patient the data required for our study. They included :

- Demographic data
- Microbiological data
- Evolving data

2.2.1. Methods

2.2.2. Biological diagnosis of co-infection

2.2.2.1. Withdrawals

A co-infection or super-infection was diagnosed when one of the following microbiological tests was positive 1 of these samples:

- Non-invasive sampling of the respiratory tract: cytobacteriological examination of sputum (CBEC). CCBT was performed after disinfection of the oral cavity with an antiseptic and during a coughing effort (assisted by kinesitherapy if necessary). Before culture, microscopic examination was performed to assess the quality of the sample. The presence of epithelial cells with a value greater than 25 cells per field indicates poor sample quality. The presence of polynuclei > 25 per field or bacteria within the polynuclei was a criterion of infection.
- Invasive respiratory sampling for mechanical ventilation.

Bronchoalveolar minilavage (BAL) was the most frequently used sample.
Blood cultures: were taken if possible before any antibiotic treatment. After disinfecting the skin with povidone-iodine, two vials of aerobic and anaerobic blood cultures were sent to the microbiology laboratory. If a fungal infection was suspected, blood cultures on fungal medium were also sent to the parasitology laboratory.

2.2.2.2. Diagnostic criteria

The diagnosis of pulmonary co-infection/superinfection was made on the basis of clinical, biological and radiological criteria, and confirmed by positive microbiological tests.
The main clinical signs of suspicion were fever, hypothermia and cough with increased sputum output. In mechanically ventilated patients, there was an increase in tracheal secretions and/or purulence. Disturbed gas exchange was also a criterion for suspicion. The presence of rales on auscultation was also taken into consideration **[37]**.
The main biological criteria for suspicion were elevated inflammatory markers: C-reactive protein (CRP) and procalcitonin, and the occurrence of polynuclear hyperleukocytosis or leukopenia.
The main radiological features included the presence of condensation and/or pleural effusion on chest imaging.
A positive microbiological result within the first 48 hours of admission was considered to be a co-infection and thereafter a superinfection.

2.2.2.3. Strain identification

Bacterial identification was carried out using conventional microbiology laboratory methods **(Table II)**: colony appearance, growth time, Gram stain, metabolic characteristics using API microgaleries (bio-Merieux, France) **(Figure 4)**.

Table II: Microbiological characteristics of the different species studied.

Bacterial species	Microbiological characteristics
Staphylococcus aureus	Gram-positive cocci grouped in clusters, mannitol-positive, producing catalase, coagulase and desoxyribonuclease
Staphylococcus a coagulase negative	Gram-positive cocci in clusters, producing catalase, without coagulase or DNase.
Enterobacteria	Gram-negative bacilli, oxidase negative, API 20E (bio-Merieux, France).
Pseudomonas aeruginosa	Gram-negative bacillus, non-fermenting, motile, strict aerobic, positive oxidase,

	pyoverdine and/or pyocyanine production, API 20NE (bio-Merieux, France).
Acinetobacter baumannii	Gram-negative, non-fermenting bacillus, strict aerobic, oxidase-negative growth in Breath Heartinfusion medium, at 44°C, API 20NE (bio-Merieux, France)
Enterococcus faecalis, Enterococcus faecium, Streptococcus spp.	Gram-positive cocci in chains, catalase negative, growth in hypersaline medium and in bile esculin medium in enterococci, api 20 Strepto (bio-Merieux, France).

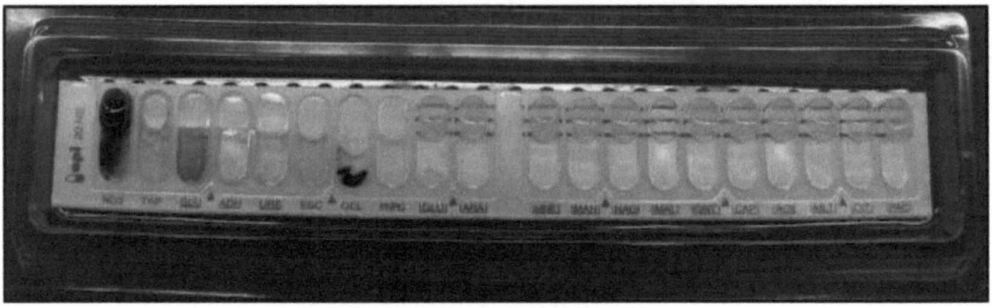

Figure 4: API 20NE gallery

2.2.3. Study of antibiotic sensitivity

Antibiotic susceptibility testing was carried out using the Mueller-Hinton gelose invasion diffusion technique in accordance with the recommendations of the Antibiogram Committee of the Societe Frangaise de Microbiologie (CA- SFM). The antibiotics tested are described in **Table III**.

Table III: Antibiotics tested during antibiograms for each species.

Species	Antibiotics tested
staphylococci	FOX: cefoxitin, PG: penicillin G, E: erythromycin, L: lincomycin, PT: pristinamycin, GN: gentamicin, TET: tetracycline, CIP: ciprofloxacin, C: chloramphenicol, AF: fusidic acid, RIF: rifampicin, SXT: sulphonamide + trimdthoprim, FOS: fosfomycin, VAN: vancomycin, TEC: teicoplanin.
enterobacteria	AMX: amoxicillin, AMC: amoxicillin + clavulanic acid, CF: cefalotin, CTX: cefotaxime, IMP: imipenem, GN: gentamicin, AN: amikacin, CIP: ciprofloxacin, SXT: sulfonamide +trimethoprim.
Pseudomonas aeruginosa	TIC: ticarcillin, TCC: ticarcillin + clavulanic acid, PIP: piperacillin, TZP: piperacillin + tazobactam, CAZ: ceftazidime, IMP: imipenem, GN: gentamicin, AN: amikacin, RIF: rifampicin, CS: colistin, CIP: ciprofloxacin.
Acinetobacter	ticarcillin, TCC: ticarcillin + clavulanic acid, PIP: piperacillin,

baumannii	TZP: piperacillin + tazobactam, CAZ: ceftazidime, IMP: imipenem, GN: gentamicin, AN: amikacin, RIF: rifampicin, CS: colistin, CIP: ciprofloxacin.
enterococci	Ampicillin, OXA: Oxacillin, GN: Gentamicin 500pg, S: Streptomycin 500pg, K: Kanamycin 1000pg, FT: Nitrofurans, C: Chloramphenicol, TET: Tetracycline, E: Erythromycin, L: Lincomycin, PT: Pristinamycin, SXT: Cotrimoxazole, RIF: Rifampicin, Fluoroquinolones, VAN: Vancomycin, TEC: Teicoplanin
streptococci	PG: Penicillin G, AMX: Amoxicillin, OXA: Oxacillin, GN: Gentamicin 500pg, S: Streptomycin 500pg, K: Kanamycin 1000 pg, TET: Tetracycline, E: Erythromycin, L: Lincomycin, PT: Pristinamycin, C: Chloramphenicol, SP: Spiramycin, SXT: Cotrimoxazole, Fluoroquinolones, RIF: Rifampicin, VAN: Vancomycin, TEC: Teicoplanin.

2.3. Statistical analysis

All the data was entered and analysed using the 20th version of IBM SPSS (Statistical Package for The Social Science) for Windows.

For qualitative variables, percentages and headcounts are calculated.

For quantitative variables, means plus or minus standard deviation were calculated. Comparisons of qualitative variables (percentages) between the periods studied were made using Pearson's Chi-square test and Fisher's exact test if one of the numbers calculated was less than 5. To compare quantitative variables (two means), we opted for Student's t-test. Differences were considered statistically significant for a p value <0.05.

CHAPTER 3

3. RESULTS

3.1. Period COVID-19

3.1.1. Epidemiological profile

3.1.1.1. Number of bacterial isolates

During the COVID-19 period, 470 bacterial isolates were isolated from bacteriological samples received from the intensive care unit. These isolates corresponded to 218 patients.

3.1.1.2. Frequency of withdrawals per month

Throughout our study period, the number of bacterial isolates varied from month to month, with values ranging from 2 to 46 and a maximum recorded in May 2021 **(Figure 5)**.

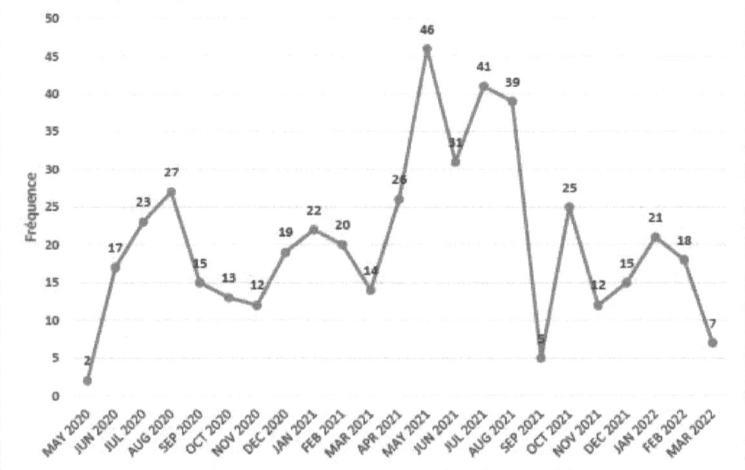

Figure 5: Monthly distribution of bacterial isolates from COVID-19 positive patients in the intensive care unit

3.1.1.3. Age

The mean age of the patients was 62.5 years, with a standard deviation of 12.5.

3.1.1.4. Gender

One hundred and thirty-eight patients were male (63.3%), giving a sex ratio (M/F) of 1.7.

3.1.2. Bacteriological profile

3.1.2.1. Distribution of isolated germs by genus and species

In our study, 470 germs were collected, 366 of which were gram-negative bacilli (GNB), i.e. a rate of 77.9%, and 104 gram-positive cocci (GPC), i.e. 22.1% of all isolates.

The BGN identified were divided into 178 non-fermentative BGN (NFGN), including 98 strains of *Acinetobacter baumannii,* 65 strains of *Pseudomonas aeruginosa* and 187 enterobacteria. The main enterobacteria were *Klebsiella pneumoniae* (78 strains) and *Escherichia coli* (45 strains).

Of the 104 isolated CGPs, staphylococci represented 92.3% (n=96), dominated by coagulase-negative staphylococci (n=59) **(Table IV)**.

Table IV: Distribution of germs isolated from COVID-19 patients in the general intensive care unit

Isolated germs	Number of strains isolated	Percentage
gram-negative bacilli	**366**	**77,9%**
Enterobacteria	**187**	**39,8%**
Klebsiella pneumoniae	78	16,7%
Escherchia coli	45	9,6%
Serratia marcescens	19	4%
Enterobacter cloacae	16	3,4%
Enterobacter aerogenes	9	1,9%
Proteus mirabilis	7	1,5%
Morganella morganii	4	0,9%
Citrobacter koseri	3	0,6%
Klebsiella oxytoca	3	0,6%
Proteus vulgaris	1	0,2%
Providencia rettgeri	1	0,2%
Providencia stuartii	1	0,2%
non-fermentative gram-negative bacilli	**178**	**37,9%**
Acinetobacter baumannii	98	20,9%
Pseudomonas aeruoginosa	65	13,8%
Stenotrophomonas Maltophilia	15	3,2%
Other Gram-negative bacilli	**1**	**0,2%**
Haemophilus influenzae	1	0,2%
gram-positive cocci	**104**	**22,1%**
Staphylococci	**96**	**20,4%**
Staphylococcus aureus	20	4,3%
Staphylococcus epidermidis	9	1,9%
Staphylococcus haemolyticus	6	1,3%
Staphylococcus hominis	2	0,4%
Other SNAs	59	12,5%
Enterococci	**7**	**1,5%**
Enterococcus faecium	4	0,9%

Enterococcus faecalis	3	0,6%
Streptococci	**1**	**0,2%**
Streptococcus pneumoniae	1	0,2%
Total	**470**	**100%**

SCN: staphylococcus a coagulase negative

3.1.2.2. Monthly distribution of bacterial isolates

During the study period, the number of germs varied from month to month. July 2021 saw a peak with a total of 37 BGN isolates identified, while CGP and BGNNF peaked in May 2021 **(Figure 6)**.

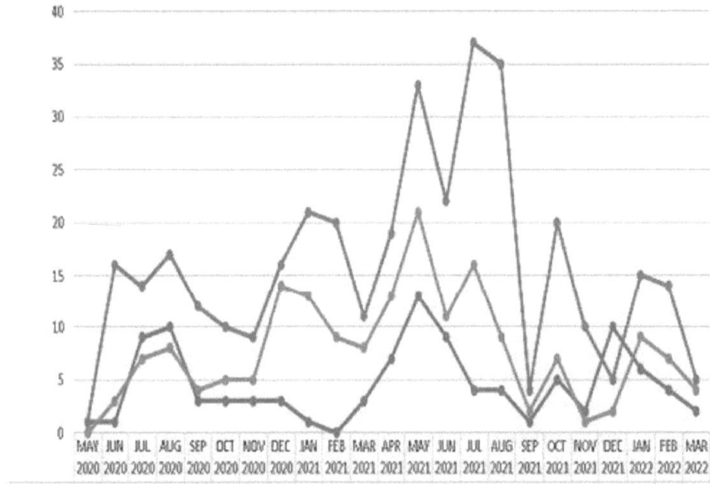

	MAY 2020	JUN 2020	JUL 2020	AUG 2020	SEP 2020	OCT 2020	NOV 2020	DEC 2020	JAN 2021	FEB 2021	MAR 2021	APR 2021	MAY 2021	JUN 2021	JUL 2021	AUG 2021	SEP 2021	OCT 2021	NOV 2021	DEC 2021	JAN 2022	FEB 2022	MAR 2022
CGP	1	1	9	10	3	3	3	3	1	0	3	7	13	9	4	4	1	5	2	10	6	4	2
BGN	1	16	14	17	12	10	9	16	21	20	11	19	33	22	37	35	4	20	10	5	15	14	5
BGNNF	0	3	7	8	4	5	5	14	13	9	8	13	21	11	16	9	2	7	1	2	9	7	4

CGP: gram-positive cocci; BGN: gram-negative bacilli; BGNNF: non-fermentative BGN

Figure 6: Monthly distribution of different groups of bacterial isolates in COVID-19 positive patients in the general intensive care unit

3.1.2.3. Nature of bacteriological samples

The bacterial isolates in our study corresponded to 418 positive samples. These were mainly tracheal aspirates (n=163; 39%), blood cultures (n=97; 23.3%), material samples (n=52; 12.4%) and urine samples (n=48; 11.5%) **(Table V)**.

Table V: Type of positive bacteriological samples taken from COVID-19 positive patients in the intensive care unit

	Number of samples	Percentage

Tracheal suction	163	39%
Blood culture	97	23,3%
Levy on equipment	52	12,4%
ECBU	48	11,5%
Spit	3	0,7%
Deeper or shallower	3	0,7%
CSF	1	0,2%
Other	51	12,2%
Total	**418**	**100%**

ECBU: cytobacteriological examination of urine; CSF: cerebrospinal fluid Other: puncture fluid, drain

❖ Distribution of bacterial strains by sampling site

In our study, 187 BGN were isolated from tracheal aspirates (51.1%) and 57 CGP were isolated from blood cultures (54.8%) **(Table VI).**

Table VI: Distribution of bacterial strains by type of sample

Type of sample	BGN (n=366)	Enterobacteria (n=187)	BGNNF (n=178)	CGP (n=104)
Tracheal suction	187 (51,1%)	69 (36,9%)	117 (65,7%)	12 (11,5%)
Blood culture	43 (11,7%)	17 (9,1%)	26 (14,6%)	57 (54,8%)
Levy on equipment	34 (9,3%)	22 (11,8%)	12 (6,8%)	23 (22,1%)
ECBU	42 (11,5%)	30 (16%)	12 (6,7%)	7 (6,7%)
Spit	4 (1,1%)	1 (0,5%)	3 (1,7%)	0 (0,0%)
Deeper or shallower	3 (1,7%)	2 (1,1%)	1 (1%)	1 (1%)
CSF	0 (0,0%)	0 (0%)	0 (0,0%)	1 (1%)
Other	53 (13,6%)	46 (24,6%)	7 (3,5%)	3 (2,9%)

ECBU: cytobacteriological examination of urine; CSF: cerebrospinal fluid; Other: puncture fluid, drain

BGN: gram-negative bacilli; BGNNF: non-fermentative gram-negative bacilli; CGP: gram-positive cocci.

The majority of enterobacteria and BGNNF were isolated from tracheal aspirates, with isolation rates of 36.9% and 65.7% respectively.

3.1.2.4. Bacterial resistance to antibiotics

- Antibiotic resistance profile in non-fermentative Gram-negative bacilli

A. baumannii strains were resistant to ceftazidime, imipenem, ciprofloxacin, amikacin and colistin in 88.8%, 88.8%, 93.9%, 51% and 0% of cases respectively. These rates were 10.8%, 16.9%, 13.9%, 7.7% and 0% respectively for *P. aeruginosa.* The rate of *A. baumannii* resistance to tigecycline was 24.5% **(Figure 7).**

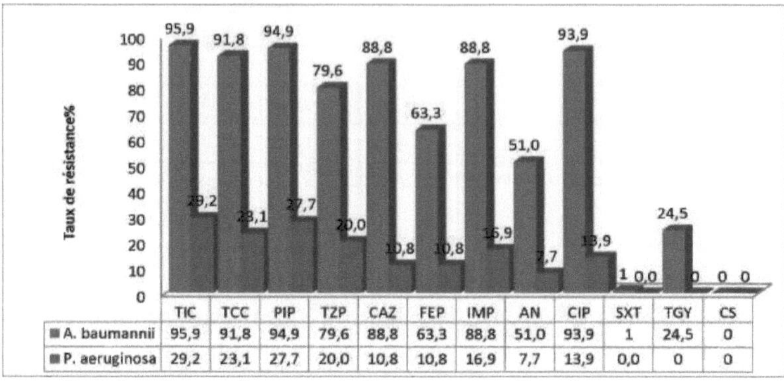

A. baumannii: Acinetobacter baumannii; P. aeruginosa: Pseudomonas aeruoginosa; TIC: Ticarcillin; TCC: Ticarcillin-clavulanic acid; PIP: Piperaciliin; TZP: Piperacilin-tazobactam; CAZ: Ceftazidime ; FEP : Cefepime ; IMP : Imipeneme ; AN: Amikacin ; CIP: Ciprofloxacin; SXT: Trimethoprime-sulfamethoxazole; TGY: Tigecycline; CS : Colistin

Figure 7: Antibiotic resistance profiles in *A. baumannii* and *P. aeruginosa*

Stenotrophomonas maltophilia strains were resistant to ticarcillin-clavulanic acid, ceftazidime, trimethoprimesulfamethoxazole and levofloxacin in 53.3%, 66.7%, 26.7% and 0% of cases respectively **(Figure 8).**

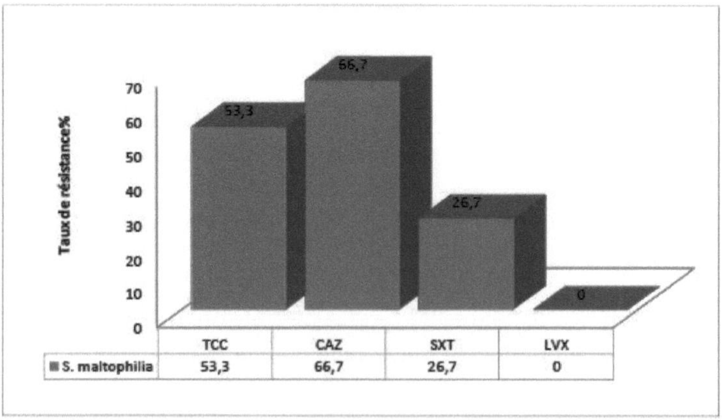

S. maltophilia: Stenotrophomonas maltophilia; TCC: Ticarcillin-clavulanic acid; CAZ:

Ceftazidime; SXT: Trimethoprim-sulfamethoxazole; LVX: Levofloxacin

Figure 8: Resistance rate of *Stenotrophomonas maltophilia* to antibiotics

- Resistance profile of enterobacteria to antibiotics

E. coli strains were resistant to cefotaxime, imipenem, ciprofloxacin, gentamicin and trimethoprimesulfamethoxazole in 35.6%, 2.2%, 68.9%, 9.3% and 68.9% of cases respectively. These rates were 42.3%, 23.1%, 51.3%, 16.1% and 50% respectively in *K. pneumoniae* **(Figure 9)**.

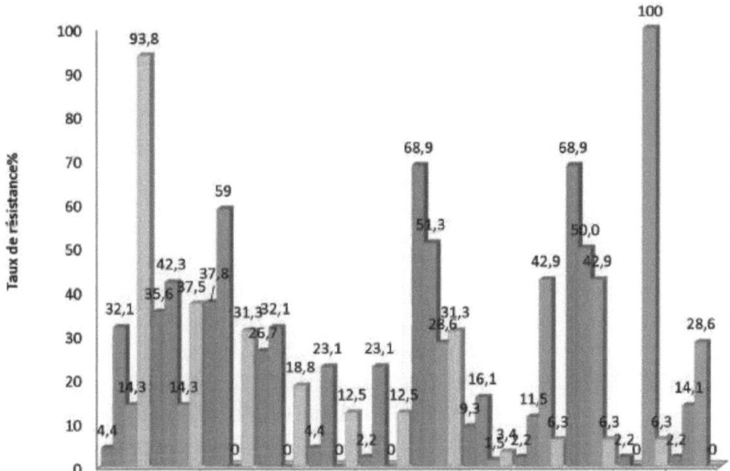

	FOX	CTX	CAZ	FEP	ETP	IMP	CIP	GM	FOS	SXT	CS	TGY
E.coli	4,4	35,6	37,8	26,7	4,4	2,2	68,9	9,3	2,2	68,9	2,2	2,2
K. pneumoniae	32,1	42,3	59	32,1	23,1	23,1	51,3	16,1	11,5	50,0	0	14,1
P. mirabilis	14,3	14,3	0	0	0	0	28,6	1,5	42,9	42,9	100	28,6
E. cloacae	93,8	37,5	31,3	18,8	12,5	12,5	31,3	3,4	6,3	6,3	6,3	0

E. coli: Escherchia coli; K. pneumoniae: Klebsiella pneumoniae; P. mirabilis: Proteus mirabilis; E. cloacae: Enterobacter cloacae; FOX: Cefoxitine; CTX: Cefotaxime; CAZ: Ceftazidime; FEP: Cefepime; ETP: Ertapeneme; IMP: Imipeneme; CIP: Ciprofloxacin; GM: Gentamicin; FOS: Fosmomycin; SXT: Trimethoprim-sulfamethoxazole; CS: Colistin; TGY: Tigecycline

Figure 9: Rates of antibiotic resistance in Enterobacteriaceae

Resistance to meticillin in *S. aureus* and SCN was 45% and 71.2% respectively.

Our *S. aureus* isolates were resistant to erythromycin, norfloxacin, fosfomycin, fusidic acid and rifamycin in 40%, 40%, 25%, 55% and 15% of cases respectively. These rates were 81.4%, 59.3%, 27.1%, 71.2% and 40.7% respectively in SCN **(Figure 10)**.

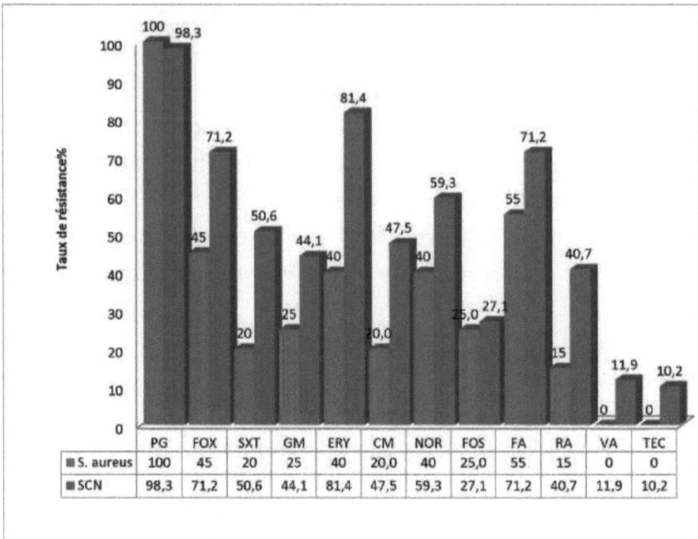

S. aureus: Staphylococcus aureus; SCN : staphylococcus a coagulase negative ; PG : Penicillin G ; FOX: Cefoxitin ; SXT: Trimethoprime-sulfamethoxazole ; GM: Gentamicin ; ERY : Erythromycin; CM: Clindamycin; NOR: Norfloxacin; FOS: Fosfomycin; FA: Fusidic acid; RA: Rifamycin; VA: Vancomycin; TEC: Teicoplanin

Figure 10: Antibiotic resistance profile of staphylococci

Our seven enterococcal isolates included four strains of *E. faecium* and three strains of *E. faecalis*. The resistance phenotypes observed are shown in **Figure 11**.

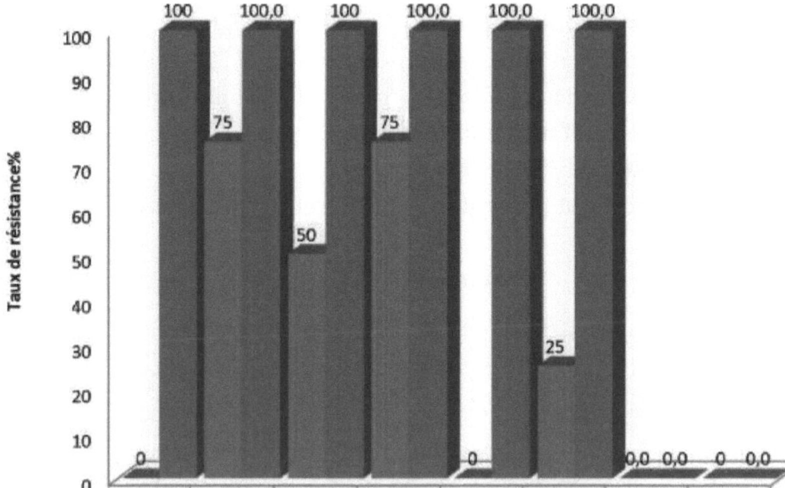

	AMP	GM	ERY	CM	PR	RA	VA	TEC
E. faecium	0	75	50	75	0	25	0,0	0
E. faecalis	100	100,0	100	100,0	100,0	100,0	0,0	0,0

E. faecium: Enterococcus faecium; E. faecalis: Enterococcus faecalis; AMP: Ampicillin; GM: Gentamicin; ERY: Erythromycin; CM: Clindamycin; PR: Pristinamycin; RA: Rifamycin; VA: Vancomycin; TEC: Teicoplanin

Figure 9: Rate of antibiotic resistance in enterococci

3.1.3. Bacterial multiresistance profile

3.1.3.1. Distribution of multi-resistant bacteria

During the study period, we identified 185 BMR out of a total of 470 isolated strains, i.e. a rate of 39.4%. Multi-resistant BGN were in the majority (n=176; 95.1%), represented mainly by multi-resistant NFNGS (n=110; 59.4%) **(Figure 12)**.

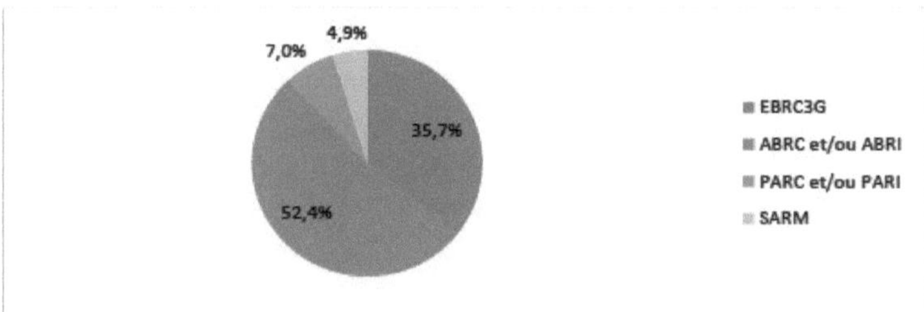

EBRC3G: enterobacteria resistant to cephalosporins of 3^{emm} generation. CARL and/or CRL: A. baumannii resistant to ceftazidime and/or imipenem. PARC and/or PARI: P. aeruginosa resistant to ceftazidime and/or imipenem SARM: S. aureus resistant to meticillin.

Figure 12: Distribution of multi-resistant bacteria

Our strains *of A. baumannii* and *P. aeruginosa* resistant to ceftazidime and/or

imipenem represented 52.4% (97 strains) and 7% (13 strains) of the total BMR, respectively. Enterobacteriaceae resistant to 3[eill] th generation cephalosporins represented 58.3% (66 strains) of our total BMR.
For the CGP, 9 strains of meticillin-resistant *S. aureus* (MRSA) (4.9%) were isolated **(Table VII)**.

Table VII: Number and percentage of different types of multi resistant

	Number of BMR strains	Percentage
BGN	176	95,1%
EBRC3G	**66**	**35,7%**
E-BLSE	60	32,4%
ERC	21	11,4%
CARL/I	**97**	**52,4%**
PARC/I	**13**	**7%**
CGP	9	4,9%
MRSA	9	4,9%
Total	185	100,0%

BMR: multi-resistant bacteria; EBRC3G: enterobacteria resistant to 3[eme] generation cephalosporins; E-BLSE: extended-spectrum beta-lactamase-producing enterobacteria; ERC: carbapenem-resistant enterobacteria; ABRC/I: A. baumannii resistant to ceftazidime and/or imipenem; PARC/I: P. aeruginosa resistant to ceftazidime and/or imipenem; MRSA: meticillin-resistant S. aureus.

12.1.3.2. Distribution of multi-resistant bacteria by type of sample

BMRs were identified mainly in 107 tracheal aspirates (57.8%) and 29 blood cultures (15.6%) **(Figure 13)**.

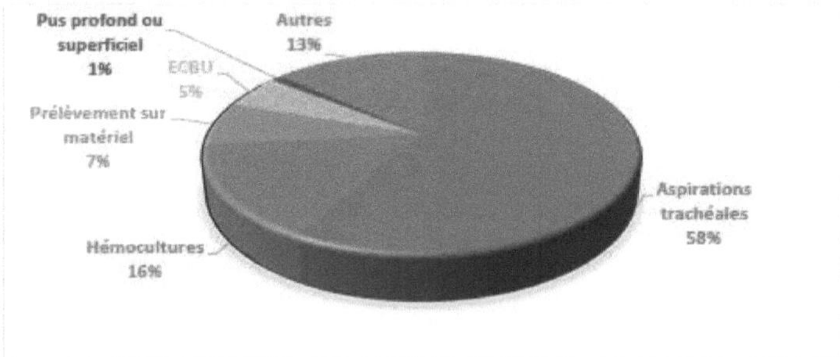

ECBU: cytobacteriological examination of urine; Other: puncture fluid, drain.

Figure 13: Distribution of multi-resistant bacteria by type of sample

13.1.4. Patient progress

Of the patients admitted to intensive care with COVID-19 infection and who developed a bacterial infection, 86.2% (n=188) died during hospitalisation **(Figure 14)**.

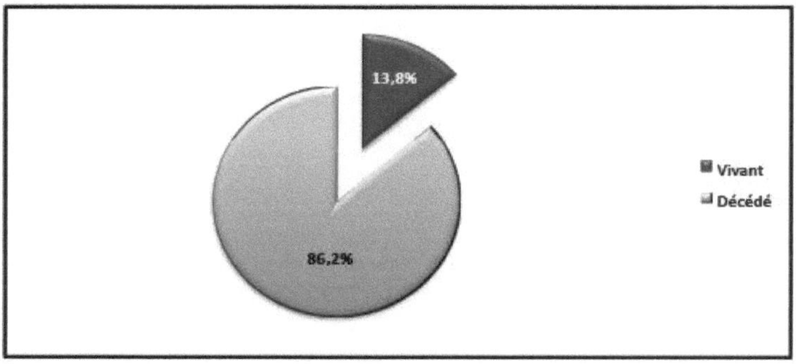

Figure 14: Patient progress during hospitalisation

3.2. Non-COVID period

3.2.1. Epidemiological profile

3.2.1.1. Age

A total of 93 patients were admitted to the general intensive care unit between 0106-2018 and 30-06-2019 for bacterial infection. The mean age of the patients was 55.6 years with a standard deviation of 17.9.

3.2.1.2. Gender

The majority of patients were male (n=62), with a percentage of 66.6% and a sex ratio of 2.

3.2.2. Bacteriological profile

3.2.2.1. Distribution of isolated germs by genus and species

In our study, 198 strains were isolated from the 93 patients included, of which 159 were BGN, representing a rate of 80.3%, and 39 were CGP, representing 19.7% of all isolates.

The BGN were divided into 79 enterobacteria, 71 BGNNF and 9 strains *of Haemophilus spp.* More than half of the enterobacteria were represented by *K. pneumoniae* (n=34; 17.2%) and *E. coli* (n=16; 8.1%).

BGNNFs were dominated *by P. aeruginosa* (n=36; 18.2%) and *A. baumannii* (n=34; 17.2%). Staphylococci (n=31; 15.7%) were the predominant PGC **(Table VIII).**

Table VIII: Type of germs isolated from patients in general intensive care during the non-COVID period

Germs		Number of strains isolated	Percentage
BGN	Enterobacteria	79	39,9%

K. pneumonia	34	17,2%
E. coli	16	8,1%
S. marcescens	5	2,5%
E. cloacae	4	2%
K. aerogenes	4	2%
M. morganii	4	2%
P. mirabilis	3	1,5%
C. koseri	3	1,5%
K. oxytoca	3	1,5%
P. stuartii	2	1%
P. rettgeri	1	0,5%
Non-fermentative BGN	71	35,9%
P. aeruginosa	36	18,2%
A. baumannii	34	17,2%
S. maltophilia	1	0,5%
Haemophilus spp	9	4,5%
H. influenza	7	3,5%
H. parainfluenzae	2	1%
Total BGN	**159**	**80,3%**
CGP **Streptococci**	2	1%
S. pneumoniae	2	1%
Staphylococci	31	15,7%
S. aureus	10	5%
SCN	21	10,6%
Enterococci	6	3%
E. faecium	1	0,5%
E. faecalis	5	2,5%
Total CGP	**39**	**19,7%**
Total germs	**198**	**100%**

K. pneumoniae: Klebsiella pneumoniae; E. coli: Escherchia coli; S. marcescens: Serratia marcescens; E. cloacae: Enterobacter cloacae; K. aerogenes: Klebsiella aerogenes
M .morganii: Morganella morganii; P. mirabilis: Proteus mirabilis; C. koseri: Citrobacter koseri; K. oxytoca: Klebsiella oxytoca; P. stuartii: Providencia stuartii; P. rettgeri: Providencia rettgeri; P. aeruginosa: Pseudomonas aeruoginosa; A. baumannii: Acinetobacter baumannii; S. maltophilia: Stenotrophomonas maltophilia; H. influenzae: Haemophilus influenzae; H. parainfluenzae: Haemophilus parainfluenzae; S. pneumoniae: Streptococcus pneumoniae; S. aureus: Staphylococcus aureus; SCN: staphylococcus a coagulase negative; E. faecium: Enterococcus faecium; E. faecalis: Enterococcus faecalis

3.2.2.2. Nature of bacteriological samples

In our series, 40.4% of strains were isolated from tracheal aspirates (n= 80; 40.4%) followed by blood cultures (n= 36; 18.2%), while strains collected from urine had an incidence of 16.7% (n= 33) **(Table IX)**.

Table IX: Type of bacteriological samples taken from patients in the intensive care unit during the non-COVID period

	Number of samples	Percentage

Tracheal suction	80	40,4%
Blood culture	36	18,2%
Levy on equipment	33	16,7%
ECBU	33	16,7%
Deeper or shallower	9	4,5%
Spit	2	1%
Other	5	2,5%
Total	**198**	**100%**

ECBU: cytobacteriological examination of urine; Other: puncture fluid, drain

3.2.2.3. Breakdown of species by sampling area

In the 80 tracheal aspirates, BGNNF were in the majority: these were essentially *P. aeruginosa* (n=21; 26.8%) and *A. baumannii* (n=16; 20%). In the 36 blood cultures taken, *SCN* were predominant (n=9; 25%) **(Figure 15)**.

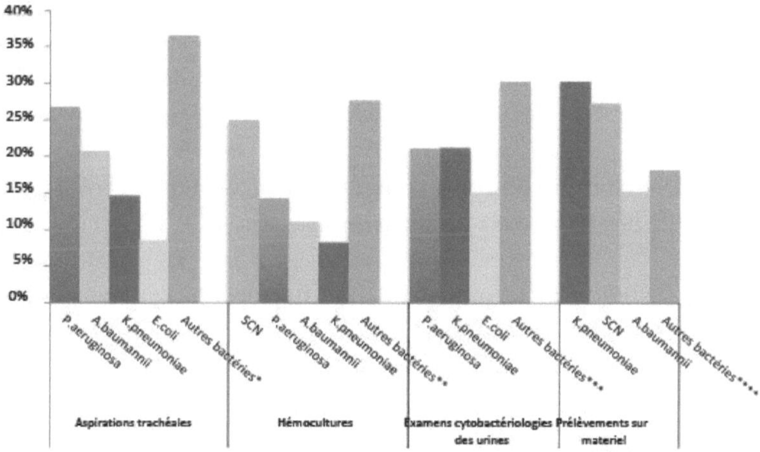

****P*. penneri, H. parainfluenzae, K. oxytoca, P. rettgeri, C. koseri, E. aerogenes, E. cloacae, E. faecalis, S. maltophilia*

***E. faecalis, E. cloacae, M. morganii, K. oxytoca, E. aerogenes, P. stuartii, E. coli, S. aureus*

****A. baumannii, E. faecalis, P. mirabilis, SCN, E. aerogenes, C. koseri*

*****P. aeruginosa, M. morganii, S. aureus*

Figure 15: Germs isolated from various samples taken from patients in the intensive care unit during the non-COVID period

3.2.3. Multi-resistant bacterial infections

During this study period, 82 BMRs were isolated, representing 41.4% of the total bacterial strains.

3.2.3.1. Distribution of different types of multi-resistant bacteria

CARL/I represented 35.5% of BMR (n=29) followed by EBRC3G (n=33; 40%). Three MRSA strains were isolated, representing 3.7% of all MRSA **(Figure 16)**.

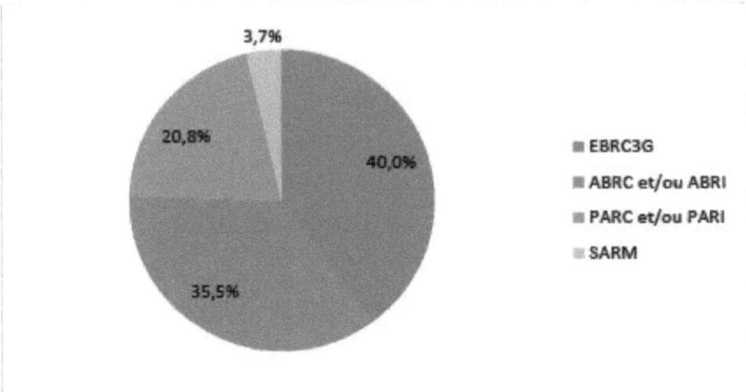

EBRC3G: Enterobacteriaceae resistant to 3rd generation cephalosporins. CARL and/or IRFA: A. baumannii resistant to ceftazidime and/or imipenem. PARC and/or PARI: ceftazidime- and/or imipenem-resistant P. aeruginosa MRSA: methicillin-resistant S. aureus.

Figure 16: Distribution of multi-resistant bacteria isolated from patients in the intensive care unit during the non-COVID period

3.2.3.2. Distribution of multi-resistant bacteria by type of sample

During this study period, almost half of the BMRs were isolated from tracheal aspirates (n=38; 46%) **(Figure 17)**.

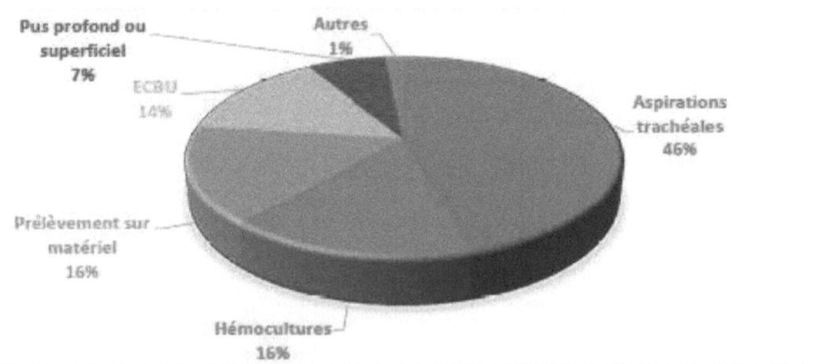

ECBU: cytobacteriological examination of urine; Other: puncture fluid, drain.

Figure 17: Distribution of multi-resistant bacteria by type of sample

3.2.4. Patient progress

More than half (51.6%) of the patients hospitalised in the general intensive care unit for bacterial infection before the COVID period had died **(Figure 18)**.

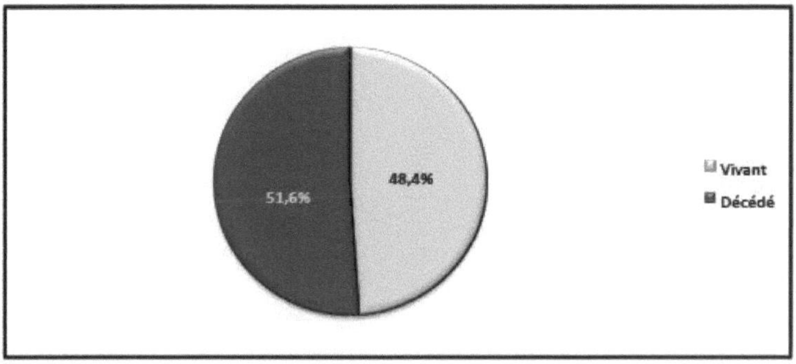

Figure 18: Trend in patients with bacterial infections during the non-COVID period

3.3. Comparative study

3.3.1. Average age

The mean ages of the patients were 62.5 and 55.6 during the COVID and non-COVID periods respectively.

The difference was statistically significant at p=0.02.

3.3.2. Gender

We noted that the male sex predominated during the COVID period as well as the non-COVID period. This predominance was not statistically significant p=0.65 **(Table X)**.

Table X: Variation in the sex of patients between the two periods

Variables		Period COVID-19 (n)	Period no COVID (n)	Total (n)	P
Gender	Men	138	62	200	0,65
	Woman	80	31	111	
	Total	218	93	311	

n: number

3.3.3. Nature of bacterial samples

There was an increase in the rate of tracheal aspirations and blood cultures during the COVID period. This increase was not statistically significant **(Figure 19)**.

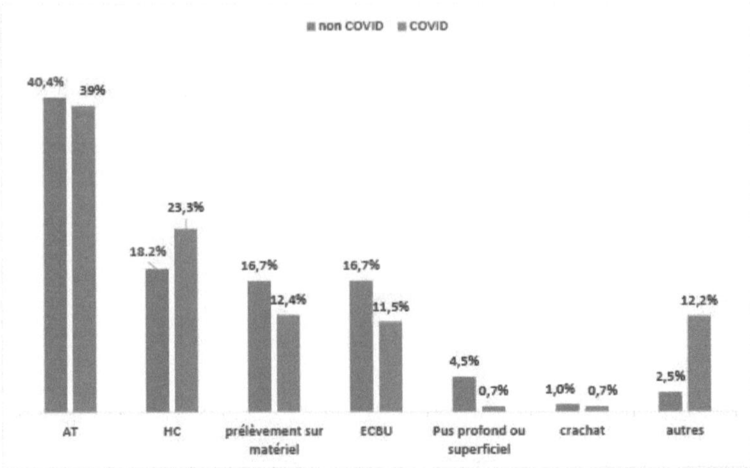

AT: tracheal aspiration; HC: blood culture; ECBU: cytobacteriological examination of urine; Other: puncture fluid, drain

Figure 19: Variation in the nature of samples over the two periods

3.3.4. Isolated germs

There was an increase in the isolation of each of: *A. baumannii, S. maltophilia, E. cloacae,* SCN, *S. marcescens* and *E. faecium* from the non-COVID period to the COVID period, unlike *K. pneumoniae and P. aeruginosa* whose isolation rates decreased slightly. These variations were not statistically significant **(Figure 20)**.

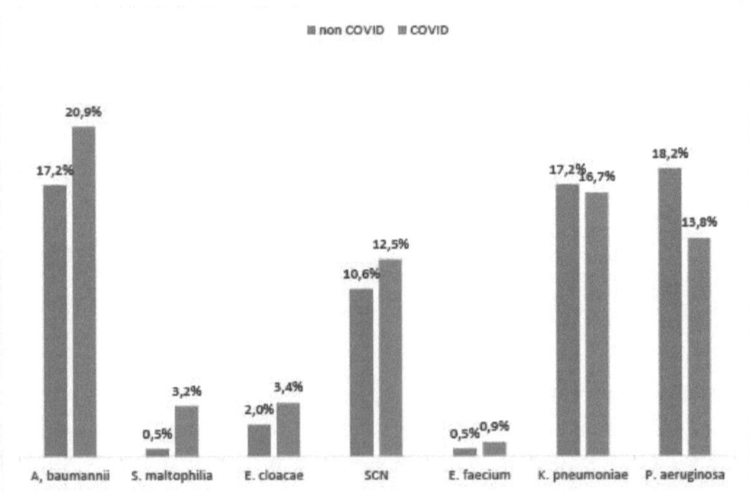

A. baumannii: *Acinetobacter baumannii*; *S. maltophilia*: *Stenotrophomonas maltophilia*; *E. cloacae*: *Enterobacter cloacae*; SCN: staphylococcus a coagulase negative; *S.*

marcescens : *Serratia marcescens; E. faecium*: *Enterococcus faecium*

Figure 20: Variation in germs over the two periods

3.3.5. Multi-resistant bacteria

Between the two periods, some BMRs rose while others fell. This increase was noted in ABRI **(Figure21).**

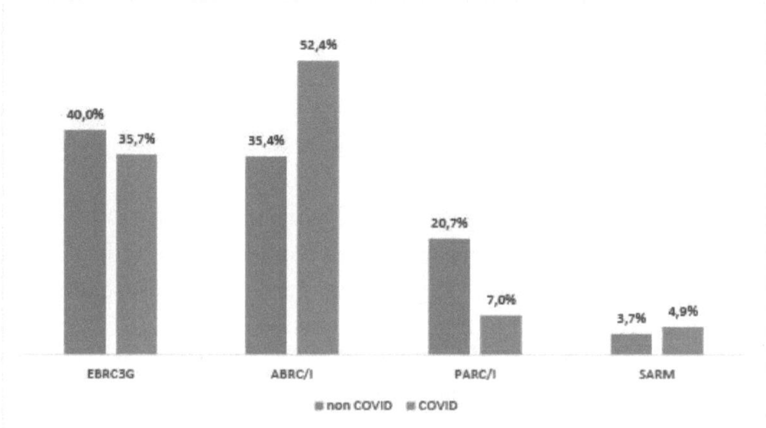

EBRC3G: enterobacteria resistant to cephalosporins of 3[eme] generation.
ABRCI: A. baumannii resistant to ceftazidime and/or imipenem.
PARCI: P. aeruginosa resistant to ceftazidime and/or imipenem
MRSA: Meticillin-resistant S. aureus

Figure 21: Variation in multi-resistant bacteria over the two periods

There was an increase in the isolation rate of CARL/I during the COVID period. This variation was statistically significant for CARL/I (p=0.03).

3.3.6. Patient progress

There was a significant increase in the death rate during the COVID period (86.2%) compared with the non-COVID period (51.6%) with p=0.02 **(Table XI)**.

Table XI: Patient outcomes during the two periods

Variables	Period COVID- 19 (n)	Non-COVID period (n)	Total (n)	P
Decedes	188	48	236	**0,02**
Living	30	45	75	
Total	218	93	311	

n: number

CHAPTER 4

4. DISCUSSION

The appearance of the SARS-CoV-2 virus was accompanied by an increase in morbidity and mortality, particularly due to bacterial infections. A large number of patients ended up being hospitalised in intensive care, under unfavourable conditions, in particular due to a healthcare system weakened by economic difficulties **[3,4,9]**. The aim of our study was to examine bacterial infections in patients hospitalised with COVID-19 in general intensive care units, and to compare them with the period preceding the COVID-19 pandemic, in particular taking into account the emergence of multi-resistant bacteria.

After reviewing medical records and analysing microbial strains, we found that factors such as male gender and age were associated with an increased risk of severity. Extensive CT involvement was closely associated with an increased risk of admission to a general intensive care unit. BGN, notably *A. baumannii*, *P. aeruginosa*, and *K. pneumoniae*, were the most frequently identified pathogens. The most common samples taken were tracheal aspirates and blood cultures. The risk of infection varied according to age and sex.

Comparing the COVID-19 period with the non-COVID-19 period, we observed a significant increase in the mean age of patients. We also observed an increased prevalence of *A. baumannii* isolation in samples, in particular ABRI

In addition, an increase in mortality was noted during the COVID-19 period compared with the non-COVID-19 period.

4.1. Study of the COVID period

4.1.1. Epidemiological profile

The majority of national studies have shown that patients admitted to intensive care units are the most likely to contract a nosocomial infection during their stay **[38]**.

According to Dridi et al, the intensive care unit presents the greatest risk of superinfection, with a rate of 23%, followed by the surgical unit with 22.4% **[39]**. A study conducted in Geneva by Hugonnet et al. also showed that intensive care units were associated with the highest risk of nosocomial infections, with a rate of 33% **[39,40]**.

4.1.1.1. Incidence

During our study period, we found that 22.9% of patients admitted to the intensive care unit with SARS-CoV-2 infection also had a concurrent bacterial infection. Our results indicate the second highest prevalence of bacterial superinfections in COVID-19 resuscitation patients, preceded by a study conducted in Saudi Arabia **(Table XII)**.

Table XII: Percentage of bacterial infections in patients COVID-19 hospitalised in medical intensive care

Studies	Percentages
Westblade LF et al (Spain, 2022) **[41].**	2,5%
Westblade LF et al (England, 2022) **[41].**	2,7%
Westblade LF et al (America, 2022) **[41].**	3,5%
Langford BJ et al (Asia, 2020) **[42].**	6,9%
Pourajam S et al (Iran, 2022) **[43].**	11,9%
Westblade LF et al (France, 2022)**[41].**	19 ,8%
Said KB et al (Saudi Arabia, 2022) **[44].**	36%
Our study	**22,9%**

The difference in results between studies may be explained by the heterogeneity of the populations studied. In addition, several authors focused solely on respiratory co-infection and did not take into account other sites of infection **[45,46]**. Poor socio-economic conditions, lack of hygiene, including limited use of personal protective equipment (PPE), reduced frequency of changing ventilation equipment and fewer qualified staff are all potential risk factors contributing to the increase in bacterial infections, according to some studies **[41]**.

4.1.1.2. Age

In our study, we found that the mean age of patients hospitalised with COVID-19 in the general intensive care unit was 62.5 years, with a standard deviation of 12.5. These results were in agreement with several other studies conducted in Germany and China, as well as a meta-analysis carried out in Asia during the same period, which also reported mean ages similar to those in our study **[43,47-49]**. This mean age continued to be recorded even in 2022, as reported by Moreno-Garda E et al, Nebreda-Mayoral T et al, Naseef HA et al, Alqahtani A et al, and Pourajam S et al **[43,46,50-52]**.

According to K.B.Said et al., bacterial co-infections in patients with COVID-19 infection were associated with an age of over 50 years **[44]**.

This observation may be explained by the combination of SARS-CoV- 2 infection with other predisposing risk factors, such as reduced immune system activity, which may make older age groups more susceptible to bacterial or fungal infections **[53]**. Older adults with COVID-19 often present with a more atypical form of the disease, with an increased risk of complications and higher mortality. In addition, the risk of death increases significantly with age **[54,55]**.

4.1.1.3. Gender

In our study, we found that 63.3% of patients were male, which is in line with the literature showing a slight male predominance **(Table XIII)**.

Table XIII: Percentage of male patients in the different studies

Studies	Percentage of males
Alqahtani A et al (Saudi Arabia, 2022) **[52].**	79,5 %
Bravo F et al (Spain, 2022) **[56]**	69%
Li J et al (China, 2020) **[47]**	66,7%
Pourajam S et al (Iran, 2022) **[43].**	63,6%
Langford BJ et al (Asia, 2020) **[42].**	54,2 %
Naseef HA et al (Palestine, 2022) **[51].**	50,5%
Our study	**63,3%**

There are several possible explanations for this disparity:

- Men have higher ACE-2 enzyme activity, regulated by male sex hormones, which may contribute to greater susceptibility to SARS-CoV-2 infection and worsening clinical outcomes **[51]**.
- Smoking rates are higher in men, which may increase their exposure to the virus and their vulnerability to respiratory infections **[51]**.
- Women may have a lower susceptibility to SARS-CoV-2 due to the role of the X chromosome and sex hormones in innate and adaptive immunity, which may partially protect them against infection **[46]**.

These hypotheses may help to understand the male dominance observed in our study and in the literature.

4.1.2. Bacteriological profile

Bacterial infections are a frequent cause of hospitalisation, and nosocomial infections are increasingly common, particularly in intensive care units where patients are particularly vulnerable to infection and often require antibiotics to improve their chances of survival.

In recent years, there has been a significant increase in the incidence of infections caused by BGN in the intensive care setting. These infections are associated with high morbidity and mortality **[57]**.

In our study, we isolated a total of 470 bacterial strains, of which 366 (77.9%) were Gram-negative bacilli (GNB) and 104 (22.1%) were Gram-positive cocci (GPC). These results are consistent with the majority of published studies. Indeed, Aurilio C et al. also found a predominance of BGN (71.4%) compared with CGP **(Table III) [58].** Studies conducted in Belgium, China and Saudi Arabia have also shown a predominance of BGN in nosocomial infections **[45,47,52]**.

The most frequently isolated organisms in our study were *A. baumannii* (20.9%), *K. pneumoniae* (16.7%), *P. aeruginosa* (13.8%) and SCN (12.5%). These results are consistent with the literature, although some differences can be attributed to patient inclusion criteria, the types of infections studied and the specific bacterial ecology of the different hospitals **(Table XIV) [43]**. For example, a study conducted in Spain showed that community co-infections were mainly caused by *S. pneumoniae* and *S. aureus*, whereas nosocomial infections were mainly caused by *P. aeruginosa* and *E. coli* **[44,45,55,59-65]**. According to Garcia-Vidal C. et al, several factors may explain the differences observed between the different studies, such as the empirical use of antibiotics, isolation measures, host macrophage activation, and the use of mechanical ventilation and catheters **[48]**. These factors may influence the distribution of bacterial strains and the resistance patterns observed in each study.

Table XIV: Germs isolated in the various studies

Studies	The most isolated germs
Said KB et al (Iran, 2022) **[44].**	*A. baumannii* (90%) *S. aureus* (10%)
Moreno-Garda E et al (Spain, 2022) **[50].**	*S. pneumoniae* (79%) *S. aureus* (6.8%) *H. influenzae* (6.8%)
Husain M et al (France, 2020) **[66]**	*P. aeruginosa* (29.6%) *S. aureus* (18.5%) *S. pneumoniae* (14.8%) *E. faecalis* (11.1%)
Li J et al (China, 2020) **[47]**	*A. baumannii* (35.8%) *K. pneumoniae* (30.8%) *S. maltophilia* (6.3%)
Westblade LF et al (Amerique, 2021) **[41].**	SCN (21%) *Enterococcus spp.* (15%) *Klebsiella spp.* (15%) *P. aeruginosa* (13%) *S. aureus* (9%)
Alqahtani A et al (Saudi Arabia, 2022) **[52].**	*K. pneumoniae* (57.69%) *A. baumannii* (11.5%)
Our study	*A. baumannii* (20.9%) *K. pneumoniae* (16.7%)

	P. aeruginosa (13.8%) SCN (12.5%)

SCN: staphylococcus a coagulase negative

4.1.2.1. Distribution of germs by sample

In our study, we found that tracheal aspirates were the most frequently positive sample, accounting for 39% of cases. Blood cultures were also frequent, with a positive rate of 23.3%.

These results are in agreement with other studies conducted by Pourajam S et al, De Bruyn A et al, and Naseef HA et al, where respiratory infections, particularly ventilator-associated pneumonia, predominated. In contrast, other studies by Zhou F et al. and Nebreda-Mayoral T et al. highlighted sepsis and urinary tract infections as the main complications in intensive care patients **[43,45,46,51,55,65,67]**.

According to the literature, respiratory complications in patients with COVID-19, such as tracheobronchitis and pneumonitis, are mainly related to reduced airway defence function following invasive procedures such as orotracheal intubation and mechanical ventilation during hospitalisation **[47,56]**.

Bacteremia was the second most frequent location of bacterial superinfections in COVID-19 intensive care patients. They were mainly due to CGP, as reported in studies conducted by de Santis V et al. and Westblade LF et al. **[41,68]**.

Blood cultures play an essential role in microbiological diagnosis, in particular differential blood cultures from venous catheters associated with venipunctures, which help to minimise erroneous results **[69]**. Data on urinary tract infections in patients with COVID-19 are limited. Most studies show a predominance of BGN such as *E. coli* and *K. pneumoniae*, with the possibility of poly-microbial infections **[44,46,47,68]**. However, a study carried out in Spain in the Valladolid region in 2022 reported a predominance of CGP **[46,48]**.

In summary, our study showed that respiratory samples were the predominant site of germ isolation, with a high prevalence of *A. baumannii*, *P. aeruginosa* and *K. pneumoniae*. Bacteremia and urinary tract infections were also frequent, with varying microbiological profiles. These results are broadly in line with the literature, although there may be variations depending on the characteristics of each population studied and the specific clinical practices of each establishment. A better understanding of the distribution of germs per sample in bacterial infections in patients admitted to intensive care units will help to guide appropriate prevention and treatment strategies to improve clinical outcomes.

4.1.2.2. Resistance of strains to antibiotics

The COVID-19 pandemic has been accompanied by an increase in the consumption of antibiotics for fear of bacterial superinfections in patients suffering from this disease. However, this increased use of antibiotics has led to the emergence and spread of antimicrobial resistance, which represents a major challenge for global health. According to a study by Costa et al, multidrug resistance was present in 96% of *A. baumannii* strains and 57% of *K. pneumoniae* strains **[70]**. It is estimated that by 2050, antimicrobial resistance could cause the death of 10 million people and incur costs of up to 100,000 billion dollars **[71]**.

- **Antibiotic resistance profile in non-fermentative Gram-negative bacilli**

In our study, almost all *A. baumannii* strains were multi-resistant bacteria, showing resistance to ceftazidime, imipenem, ciprofloxacin, amikacin and colistin in 88.8%, 88.8%, 93.9%, 51% and 0% of cases respectively. These results were similar to those reported by Alqahtani A et al, Pourajam S et al, Said KB et al and Nebreda Mayoral T et al **[43,44,46,47,52]**.

According to Perez S et al, the increase in the isolation of multi-resistant *A. baumannii* in healthcare establishments can be attributed to its ability to easily contaminate the patient care environment and the hands of healthcare staff, as well as its ability to survive for long periods on dry surfaces and to be transmitted by colonised asymptomatic individuals **[72]**. These factors make this epidemic difficult to control in intensive care units **[72]**.

A report on patients with COVID-19 in 19 intensive care units in China showed that imipenem-resistant *A. baumannii* was identified in 63% of isolates associated with secondary infections, mainly ventilator-associated pneumonia **[70]**.

For *P. aeruginosa*, the resistance rates observed in our study for imipenem, ceftazidime, amikacin and colistin were 16.9%, 10.8%, 7.7% and 0% respectively. These results are consistent with those reported by Said KB et al, with the exception of colistin resistance which was absent in our study (6.7% in Said KB et al) **[44]**.

- **Resistance profile of enterobacteria to antibiotics**

Enterobacteriaceae, which are Gram-negative bacilli, are one of the most important families of bacteria involved in human disease. They include both commensal bacteria found mainly in the digestive tract **[73]** and strict pathogens such as *Salmonella Typhi* and *Shigella dysenteriae*. Other opportunistic enterobacteria can cause infections, mainly in intensive care settings, in patients who are often frail. The acquired resistance of enterobacteria to aminopenicillins is high, but this resistance decreases in the presence of clavulanic acid. The combination of piperacillin and tazobactam proved to be the most effective. Similar results were observed in Cameroon and Spain **(74-76)**.

In our study, we found that our enterobacteria showed resistance to third-generation cephalosporins in 35.3% of cases. This resistance was first described in 1984 in Germany **[77]** and has been observed at the Charles Nicolle Hospital in Tunis since 1985 **[78]**, with a first report in Sfax in 1988 **[79]**. Significant variations have been observed depending on the bacterial species. In our series, the highest rate of resistance to third-generation cephalosporins was observed in *K. pneumoniae* (89.1%). This resistance may be attributed either to the acquisition of an extended-spectrum betalactamase, or to the acquisition of a cephalosporinase of chromosomal or plasma origin **[80]**.

In comparison with other recent studies, we observed a significantly higher rate of resistance to third-generation cephalosporins in the KES group (*Klebsiella*, *Enterobacter* and *Serratia*) compared with Europe and the United States, where the rate did not exceed 5% **[81,82]**. This difference could be explained by the prevalence of nosocomial infections, the effect of antibiotic selection pressure linked to certain healthcare practices, and the failure to comply with hygiene measures in the presence of carriers of multi-resistant bacteria. Carbapenems have been shown to be the most effective beta-lactam antibiotics against enterobacteria, both in our study and in the literature **[83]**.

- **Resistance profile of gram-positive cocci to antibiotics**

PFMCs, including *S. aureus*, coagulase-negative staphylococci and *Enterococcus spp.* are pathogens commonly implicated in hospital-acquired infections worldwide. One of the major concerns associated with these micro-organisms is their increasing resistance to antibiotics, which can lead to therapeutic failure. Combating antibiotic resistance in PFMC is therefore essential to ensure effective treatment and reduce the complications associated with nosocomial infections **[84]**.

In recent years, there has been a marked increase in the incidence of multidrug resistance in Gram-positive cocci in the intensive care setting **[85]**. In our study, the majority of strains studied showed resistance to penicillin G, which is in agreement with the results of a Chinese study **[86]** and an African series **[87,88]**. This rate of resistance is higher than that reported by the Laboratoire d'Antibiotherapie et de Recherche Therapeutique (LART) and other similar series. Resistance to meticillin was observed in 45% of Staphylococcus aureus strains in our study. This result is in agreement with a Cameroonian study **[87]**, but differs from the Algerian and Egyptian studies **[89]** where the resistance rate reached 90% **[90]**.

Aminoglycosides have been frequently used in clinical practice, particularly in combination with betalactam antibiotics or glycopeptides, for the treatment of life-threatening infections such as staphylococcal endocarditis. In our study, we observed a gentamicin resistance rate of 25% in S. aureus, which is similar to the results of a previous national series **[91]**. However, it should be noted that a study carried out in Ethiopia reported a lower rate of resistance, which may be explained by the unavailability of aminoglycosides **[92]**.

4.1.3. Multi-resistant bacteria in ventilated patients

The spread of antibiotic-resistant bacteria is a major public health problem worldwide **[93]**. Infections caused by these bacteria are often complex to treat, leading to longer hospital stays, increased mortality and higher resource costs in terms of time, labour and consumables used **[94,95]**.

In ventilated patients, tracheal aspirations were the most BMR-rich sampling site. According to an Italian study, this was attributed to the long duration of mechanical ventilation and prolonged stay in intensive care units **[68]**.

Among isolated BMRs, *A. baumannii* was the most frequently found pathogen, accounting for 52.4% of cases, which is consistent with the results of the study conducted by Westblade LF et al. However, according to another study conducted by Said KB et al, *P. aeruginosa* was the pathogen most frequently responsible for ventilator-associated pneumonia, accounting for 34.7% of cases **[44]**.

4.1.4. Patient progress

In our study, we observed that among patients admitted to intensive care for COVID-19 infection associated with bacterial infection, a high mortality rate was recorded, reaching 86.2%. These results are consistent with many other studies which have also reported a high number of deaths in similar situations **[41,43,44,46,47,55,59-64,96]**.

According to the study carried out by Gomez-Belda AB et al, the rate of COVID-19 letalite increases significantly with age. The results showed that the rate of letalite was less than 1% in people under the age of 50, increasing to 1.3% in people aged 50 and 3.6% in people aged 60 **[54]**.

Despite the strict containment and quarantine measures put in place worldwide, mortality from COVID-19 has continued to rise, as several studies have shown **[68,97]**.

According to an Italian study conducted in March 2020, it was concluded that patients with bacteremia or who developed a pulmonary infection had a higher risk of death in the intensive care unit **[68]**. These results suggest that bacterial infections may be a major cause of the high mortality rate in patients with COVID-19, as other studies have also shown **[52,98]**.

According to a Spanish study carried out in February 2022, several factors have been identified as being predictive of mortality in patients with COVID-19. These factors included male sex with various comorbidities, the presence of pneumonia, bacteremia and superinfections by *A. baumannii* **[46,47]**. Other factors may also contribute, such as advanced age and living conditions, which may be associated with a higher rate of antimicrobial resistance **[51]**. These findings highlight the importance of taking these risk factors into account when assessing and managing patients with COVID-19, in order to improve clinical outcomes and reduce mortality.

4.2. Study of the non-COVID period

Bacterial infections in the intensive care setting have always been a major public health challenge and a frequent cause of hospitalisation, even before the COVID-19 era. Over the last 15 to 20 years, this situation has worsened due to the emergence of bacterial resistance and the limited availability of effective antibiotics **[50]**. This combination of factors has created an environment conducive to the development and spread of infections that are difficult to treat, thereby increasing the complexity and severity of cases encountered in the intensive care unit. Appropriate prevention and management measures are essential to meet this challenge and reduce the impact of bacterial infections in the intensive care setting.

4.2.1. Epidemiological profile

Every year, more than 650,000 people in the United States are affected by hospital-acquired bacterial infections, and more than 20% of these infections are caused by antimicrobial-resistant pathogens **[99]**.

According to the study by Klein EY et al, advanced age and male gender were identified as major risk factors for bacterial infections. These results were confirmed by our own study **[100]**.

The main site of sampling was respiratory, in our series as well as the majority of studies, followed by blood cultures and urine. However, according to MacVane et al, the most common bacterial infections in intensive care units were pneumonia, surgical site infection, followed by gastrointestinal infections, urinary tract infections and bacteremia. According to a 2017 surveillance report on 11,282 patients from 183 US hospitals, an estimated *39%* of pneumonia cases were associated with mechanical ventilation. This form of pneumonia, known as mechanical ventilation-associated pneumonia, therefore represents a significant proportion of nosocomial lung infections. In addition, influenza causes major epidemics every year, affecting up to 20% of the population and causing considerable morbidity and mortality. These figures highlight the scale of the problem and underline the importance of implementing prevention and control measures to limit the spread of these infections and reduce their impact on public health **[99]**. Bacterial pathogens were a major cause of this morbidity and mortality **[99]**.

4.2.2. Bacteriological profile

The prevalence of bacterial infections in the intensive care setting caused by gram-negative bacilli (GNBs) was higher than that of gram-positive cocci (GPCs), which were decreasing. This trend can be explained in part by the development of new antimicrobials aimed primarily at the control and treatment of infections caused by GPC **[99]**. These results are consistent with our study, in which Gram-negative bacilli (GNB) represented 80.3% of isolates, with a predominance of *P. aeruginosa* (18.2%). They are also in line with an international study of 4,947 patients from 75 countries, where 62% of positive isolates were BGN. The most frequently reported species were *Pseudomonas spp* (20%) and *E. coli* (16%), followed by *Klebsiella spp* and *Enterobacter spp*. These data underline the importance of BGN as the predominant pathogens of bacterial infections in the ICU setting worldwide **[99]**.

In our study, EBRC3G was the most isolated type of MRB, accounting for 40%, followed by ABRC/I for 35.5%, while MRSA represented only 3.7% of MRBs.

A study conducted by Sahloul University Hospital in Sousse, Tunisia, over a period of 6 years (2012-2018) examined cases of bacteremia in the intensive care setting. The main bacteria isolated were *A. baumannii* (26.7%), *K. pneumoniae* (20.5%) and *P. aeruginosa* (9.6%). A large proportion of the isolated *A. baumannii* strains were resistant to imipenem (92.4%), but no resistance to colistin was recorded. For *K. pneumoniae*, resistance to imipenem was present in 4.1% of cases. For *P. aeruginosa*, resistance to ceftazidime was observed in 21.5% of cases and resistance to imipenem in

30.4% of cases **[101]**.

According to a study carried out in 2019 in Mumbai, India, 686 cases of bacteremia were recorded in the intensive care unit of a private hospital. 80% of the germs isolated were BGN and 10% were CGP. The prevalence of ESBL producers was 80%. Carbapenem resistance rates averaged 50% **[102]**.

A study carried out in 2017 in the central intensive care unit of the Kragujevac clinical centre in Serbia revealed that the most common pathogens were *Acinetobacter spp.* and *P. aeruginosa*, accounting for more than 60% of isolates. The rate of susceptibility to carbapenems and 4th generation cephalosporins was around 30-35% for *P. aeruginosa* and around 15-20% for *A. baumannii*. No resistance was observed for colistin and tigecycline against Gram-negative bacilli. In the case of methicillin-resistant *S. aureus* (MRSA), no resistance to glycopeptides or linezolid has been detected **[103]**.

According to a Tunisian study carried out in 2018, non-fermentative pathogens, in particular *A. baumannii*, were the most frequently identified microorganisms in intensive care units, accounting for 66% of cases. A large proportion of these bacteria were resistant to carbapenems, mainly due to the production of the OXA-23 enzyme. Enterobacteriaceae were the second most frequently isolated micro-organisms, and there was an emergence of enterobacteriaceae producing extended-spectrum beta-lactamases (ESBLs) and carbapenemases **[104]**.

4.2.3. Patient progress

In our study, we found that more than half (51.6%) of patients admitted to intensive care with a bacterial infection before the COVID-19 period unfortunately died. These results are similar to those of the study conducted by Solmaz i et al, which also recorded a mortality rate of 45.9% **[105]**.

Kumari P et al recorded a mortality rate of 69.6% in ventilated patients and 34.1% in non-ventilated patients **[106]**. These figures highlight the seriousness of bacterial infections in the intensive care setting and the importance of taking effective measures to prevent and treat these infections, in order to reduce the associated morbidity and mortality.

4.3. Comparative study of the two periods

According to a Turkish study published in March 2022, it was found that before the pandemic, around 7.5% of patients hospitalised in the general intensive care unit had bacterial infections. However, this percentage increased to 11.1% of patients hospitalised in the COVID-19 intensive care unit after the pandemic. Mortality also increased, from 83.6% to 94%

between the pre-COVID and COVID-19 periods **[107]**. These results are consistent with the findings of our study, highlighting the significant impact of the pandemic on bacterial infections and mortality in the intensive care setting.

It is interesting to note that the germs isolated before and after the COVID-19 period were similar. According to the study conducted by Bahge YG et al, the most frequently isolated micro-organisms in pre-pandemic intensive care units were *A. baumannii, K. pneumoniae, P. aeruginosa, S. aureus* and *E. coli*. During the COVID-19 period, the most frequently detected micro-organisms were *A. baumannii, K. pneumoniae, P. aeruginosa* and *E. faecium* **[107]**. These results are consistent with our study, where we mainly isolated *A. baumannii, P. aeruginosa, K. pneumoniae* and SCN during the COVID-19 period. However, during the non-COVID period, *P. aeruginosa* predominated, followed by A. *baumannii* and *K. pneumoniae*. These data underline the importance of continuous surveillance for pathogens in the intensive care setting, particularly during the COVID-19 pandemic period.

According to Bahge YG et al, antibiotic resistance rates in *A. baumannii* strains increased after the pandemic, with the exception of tigecycline **[107]**. This differs from our study, in that a rise in resistance to tigecycline was isolated. For *K.pneumoniae*, an increase in resistance to meropenem, ertapenem, amoxicillin-clavulanic acid, piperacillin-tazobactam combination, ciprofloxacin, tigecycline and cefepime was noted **[107]**. This was confirmed by our study.

CONCLUSION

Bacterial infections have always been a major public health problem worldwide, particularly in view of the worrying spread of multi-drug resistance to antibiotics. This problem has been further exacerbated by the COVID-19 pandemic, which has highlighted the shortcomings of healthcare systems in many areas, while also highlighting the ability of nations to meet the challenges through cooperation between different players. In Tunisia, the situation is particularly dramatic due to the economic crisis and ageing infrastructure, which has resulted in an alarming 29,423 deaths since the start of 2022.

Our study is one of the first to examine the impact of SARS-CoV-2 infection on the epidemiology of bacterial infections in Tunisian intensive care settings, comparing these results with the pre-pandemic period.

After studying 470 medical records and bacterial strains, we observed that advanced age and male sex were risk factors for bacterial infections in patients admitted to the intensive care unit for COVID-19.

We also observed that BGN were the most frequently isolated microorganisms, in particular *A. baumannii*, which showed a significant increase compared with the pre-COVID period. These strains were predominantly CARL/I and were mainly found in tracheal aspirate samples. This was one of the main determinants of severity and mortality, especially in view of the resistance noted to certain antibiotics.

It is essential that healthcare staff maintain rigorous standards of hand hygiene and standard and specific precautions in order to prevent infections and limit the spread of multi-resistant bacteria. In addition, it is crucial that antibiotics are prescribed judiciously to reduce the selection of resistant strains. This approach is essential to prevent loss of life, inefficient use of antibiotics and prolonged hospital stays in intensive care. It is generally recommended that appropriate antimicrobial therapy should be initiated rapidly as soon as an infection is diagnosed in critically ill patients.

One of the greatest challenges posed by suspected multi-resistant bacterial infections is to rationalise the prescription of broad-spectrum treatments to cover the most likely pathogens and strike a balance between efficacy and collateral damage.

Local surveillance is strongly recommended, as resistance patterns may vary from one ICU to another. Each ICU has its own characteristics, depending on the country, the local antibiotic resistance profile, the state of health of the

patients, their specific characteristics and the severity of the infections.
According to Martin-Loeches I et al, it is recommended to adopt alternative dosing strategies based on the principles of pharmacokinetics and pharmacodynamics rather than following a standard dosage. This approach optimises the efficacy of antimicrobial treatment and improves clinical outcomes. By adjusting the dosage according to the specific characteristics of the patient, such as state of health, renal function, body weight and other factors, it is possible to achieve optimal therapeutic concentrations for combating bacterial infections. This helps reduce the risk of resistance developing and maximises the effectiveness of the treatment.
Given the scale of the problem posed by COVID-19 and the vital risk posed by this topical issue, research into the emergence of new bacterial resistance mechanisms represents a major challenge, especially given the limited number of antibiotics available.

REFERENCES

1. Shah NS, Greenberg JA, McNulty MC, Gregg KS, Riddell J, Mangino JE, et al. Bacterial and viral co-infections complicating severe influenza: Incidence and impact among 507 U.S. patients, 2013-14. J Clin Virol. 2016;80:12-9.

2. Rice TW, Rubinson L, Uyeki TM, Vaughn FL, John BB, Miller RR, et al. Critical illness from 2009 pandemic influenza A virus and bacterial coinfection in the United States. Crit Care Med. 2012;40(5):1487-98.

3. World Health Organization. Clinical management of COVID-19: interim guidance. Geneva: WHO; 2020.

4. Alhazzani W, M0ller MH, Arabi YM, Loeb M, Gong MN, Fan E, et al. Surviving Sepsis Campaign: guidelines on the management of critically ill adults with Coronavirus Disease 2019 (COVID-19). Intensive Care Med. 2020;46(5):854-87.

5. Elmezayen AD, Al-Obaidi A, §ahin AT, Yelekgi K. Drug repurposing for coronavirus (COVID-19): in silico screening of known drugs against coronavirus 3CL hydrolase and protease enzymes. J Biomol Struct Dyn. 2021;39(8):2980-92.

6. Sarma P, Shekhar N, Prajapat M, Avti P, Kaur H, Kumar S, et al. In-silico homology assisted identification of inhibitor of RNA binding against 2019-nCoV N-protein (N terminal domain). J Biomol Struct Dyn. 2021;39(8):2724-32.

7. World Health Organization. WHO coronavirus (COVID-19) dashboard [Online]. 2021 [Accessed 30/11/2021]. Available from: https:// covid19.who.int

8. Elfiky AA. Natural products may interfere with SARS-CoV-2 attachment to the host cell. J Biomol Struct Dyn. 2021;39(9):3194-203.

9. Chen T, Wu D, Chen H, Yan W, Yang D, Chen G, et al. Clinical characteristics of 113 deceased patients with coronavirus disease 2019: retrospective study. BMJ. 2020;368:1-12.

10. Tyrrell D, Bynoe ML. Cultivation of viruses from a high proportion of patients with colds. Lancet. 1966;287:76-7.

11. Chen X, Ran L, Liu Q, Hu Q, Du X, Tan X. Hand hygiene, mask-wearing behaviors and its associated factors during the COVID-19 epidemic: A cross-sectional study among primary school students in Wuhan, China. Int J Environ Res Public Health. 2020;17(8):1-11.

12. Ravi V, Saxena S, Panda PS. Basic virology of SARS-CoV 2. Indian J

Med Microbiol. 2022;40(2):182-6.
13. Shereen MA, Khan S, Kazmi A, Bashir N, Siddique R. COVID-19 infection: Emergence, transmission, and characteristics of human coronaviruses. J Adv Res. 2020;24:91-8.
14. World Health Organization. Modes of transmission of virus causing COVID-19: implications for IPC precaution recommendations. Geneva: WHO; 2020.
15. Salas-Benito JS, De Nova-Ocampo M. Viral interference and persistence in mosquito-borne flaviviruses. J Immunol Res. 2015;2015:1-15.
16. Makoti P, Fielding BC. HIV and Human coronavirus coinfections: A historical perspective. Viruses. 2020;12(9):1-13.
17. Griffiths EC, Pedersen AB, Fenton A, Petchey OL. The nature and consequences of coinfection in humans. J Infect. 2011;63(3):200-6.
18. Van der Hoek L, Pyrc K, Berkhout B. Human coronavirus NL63, a new respiratory virus. FEMS Microbiol Rev. 2006;30(5):760-73.
19. Bengoechea JA, Bamford CG. SARS-CoV-2, bacterial co-infections, and AMR: the deadly trio in COVID-19? EMBO Mol Med. 2020;12(7):1-4.
20. Hoque MN, Akter S, Mishu ID, Islam MR, Rahman MS, Akhter M, et al. Microbial co-infections in COVID-19: Associated microbiota and underlying mechanisms of pathogenesis. Microb Pathog. 2021;156:1-10.
21. Nepogodiev D, Bhangu A, Glasbey JC, Li E, Omar OM, Simoes JF, et al. Mortality and pulmonary complications in patients undergoing surgery with perioperative SARS-CoV-2 infection: an international cohort study. Lancet. 2020;396:27-38.
22. Hendaus MA, Jomha FA, Alhammadi AH. Virus-induced secondary bacterial infection: a concise review. Ther Clin Risk Manag. 2015;11:1265-71.
23. Hendaus MA, Jomha FA. COVID-19 induced superimposed bacterial infection. J Biomol Struct Dyn. 2021;39(11):4185-91.
24. Rose MC, Voynow JA. Respiratory tract mucin genes and mucin glycoproteins in health and disease. Physiol Rev. 2006;86(1):245-78.
25. Akira S, Uematsu S, Takeuchi O. Pathogen recognition and innate immunity. Cell. 2006;124(4):783-801.
26. Chamaillard M, Hashimoto M, Horie Y, Masumoto J, Qiu S, Saab L, et al. An essential role for NOD1 in host recognition of bacterial peptidoglycan containing diaminopimelic acid. Nat Immunol. 2003;4(7):702-7.
27. Cook DN, Pisetsky DS, Schwartz DA. Toll-like receptors in the pathogenesis of human disease. Nat Immunol. 2004;5(10):975-9.

28. Doughty LA, Nguyen KB, Durbin JE, Biron CA. A Role for IFN-aв in virus infection-induced sensitization to endotoxin. J Immunol. 2001;166(4):2658- 64.
29. Passariello C, Schippa S, Conti C, Russo P, Poggiali F, Garaci E, et al. Rhinoviruses promote internalisation of Staphylococcus aureus into non-fully permissive cultured pneumocytes. Microbes Infect. 2006;8(3):758-66.
30. Kim D, Quinn J, Pinsky B, Shah NH, Brown I. Rates of co-infection between SARS-CoV-2 and other respiratory pathogens. JAMA. 2020;323(20):2085- 6.
31 World Health Organization. Coronavirus disease (COVID-19). Geneva: WHO; 2020.
32. Hendaus MA. Remdesivir in the treatment of coronavirus disease 2019 (COVID-19): a simplified summary. J Biomol Struct Dyn. 2021;39(10):3787-92.
33. Zhou F, Yu T, Du R, Fan G, Liu Y, Liu Z, et al. Clinical course and risk factors for mortality of adult inpatients with COVID-19 in Wuhan, China: a retrospective cohort study. Lancet Lond Engl. 2020;395:1054-62.
34. Guan W, Ni Z, Hu Y, Liang W, Ou C, He J, et al. Clinical characteristics of coronavirus disease 2019 in China. N Engl J Med. 2020;382(18): 1708-20.
35. Chen N, Zhou M, Dong X, Qu J, Gong F, Han Y, et al. Epidemiological and clinical characteristics of 99 cases of 2019 novel coronavirus pneumonia in Wuhan, China: a descriptive study. Lancet. 2020;395:507-13.
36. Meier MA, Branche A, Neeser OL, Wirz Y, Haubitz S, Bouadma L, et al. Procalcitonin-guided antibiotic treatment in patients with positive blood cultures: A patient-level meta-analysis of randomized trials. Clin Infect Dis. 2019;69(3):388-96.
37. Lansbury L, Lim B, Baskaran V, Lim WS. Co-infections in people with COVID-19: a systematic review and meta-analysis. J Infect. 2020;81(2):266-75.
38. Kallel H, Bahoul M, Ksibi H, Dammak H, Chelly H, Hamida CB, et al. Prevalence of hospital-acquired infection in a Tunisian hospital. J Hosp Infect. 2005;59(4):343-7.
39. Dridi E, Chetoui A, Zaoui A. Investigation of the prevalence of nosocomial infection in a Tunisian regional hospital. Sante Publique Vandoeuvre-- Nancy Fr. 2006;18(2):187-94.
40. Hugonnet S, Pittet D. Hand hygiene revisited: Lessons from the past and present. Curr Infect Dis Rep. 2000;2(6):484-9.
41. Westblade LF, Simon MS, Satlin MJ. Bacterial coinfections in

coronavirus disease 2019. Trends Microbiol. 2021;29(10):930-41.
42. Langford BJ, So M, Raybardhan S, Leung V, Westwood D, MacFadden DR, et al. Bacterial co-infection and secondary infection in patients with COVID- 19: a living rapid review and meta-analysis. Clin Microbiol Infect. 2020;26(12):1622-9.
43. Pourajam S, Kalantari E, Talebzadeh H, Mellali H, Sami R, Soltaninejad F, et al. Secondary bacterial infection and clinical characteristics in patients with COVID-19 Admitted to two intensive care units of an academic hospital in Iran during the first wave of the pandemic. Front Cell Infect Microbiol. 2022;12:1-9.
44. Said KB, Alsolami A, Moussa S, Alfouzan F, Bashir AI, Rashidi M, et al. COVID-19 clinical profiles and fatality rates in hospitalized patients reveal case aggravation and selective co-infection by limited gram-negative bacteria. Int J Environ Res Public Health. 2022;19(9):1-16.
45. De Bruyn A, Verellen S, Bruckers L, Geebelen L, Callebaut I, De Pauw I, et al. Secondary infection in COVID-19 critically ill patients: A retrospective single-center evaluation. BMC Infect Dis. 2022;22(1):1-7.
46. Nebreda-Mayoral T, Miguel-Gomez MA, March-Rossello GA, Puente-Fuertes L, Canton-Benito E, Martinez-Garcia AM, et al. Bacterial/fungal infection in hospitalized patients with COVID-19 in a tertiary hospital in the community of Castilla y Leon, Spain. Enfermedades Infecc Microbiol Clin. 2022;40(4):158-65.
47. Li J, Wang J, Yang Y, Cai P, Cao J, Cai X, et al. Etiology and antimicrobial resistance of secondary bacterial infections in patients hospitalized with COVID-19 in Wuhan, China: a retrospective analysis. Antimicrob Resist Infect Control. 2020;9(1):1-7.
48. Garcia-Vidal C, Sanjuan G, Moreno-Garda E, Puerta-Alcalde P, Garcia-Pouton N, Chumbita M, et al. Incidence of co-infections and superinfections in hospitalized patients with COVID-19: a retrospective cohort study. Clin Microbiol Infect. 2021;27(1):83-8.
49. Rothe K, Feihl S, Schneider J, Wallnofer F, Wurst M, Lukas M, et al. Rates of bacterial co-infections and antimicrobial use in COVID-19 patients: A retrospective cohort study in light of antibiotic stewardship. Eur J Clin Microbiol Infect Dis. 2021;40(4):859-69.
50. Moreno-Garda E, Puerta-Alcalde P, Letona L, Meira F, Duenas G, Chumbita M, et al. Bacterial co-infection at hospital admission in patients with COVID-19. Int J Infect Dis. 2022;118:197-202.
51. Naseef HA, Mohammad U, Al-Shami N, Sahoury Y, Abukhalil AD,

Dreidi M, et al. Bacterial and fungal co-infections among ICU COVID-19 hospitalized patients in a Palestinian hospital: a retrospective cross-sectional study. F1000Research. 2022;11:30.
52. Alqahtani A, Alamer E, Mir M, Alasmari A, Alshahrani MM, Asiri M, et al. Bacterial coinfections increase mortality of severely ill COVID-19 patients in Saudi Arabia. Int J Environ Res Public Health. 2022;19(4):1-11.
53. Rafat Z, Ramandi A, Khaki PA, Ansari S, Ghaderkhani S, Haidar H, et al. Fungal and bacterial co-infections of the respiratory tract among patients with COVID-19 hospitalized in intensive care units. Gene Rep. 2022;27:1- 5.
54. Gomez-Belda AB, Fernandez-Garces M, Mateo-Sanchis E, Madrazo M, Carmona M, Piles-Roger L, et al. COVID-19 in older adults: What are the differences with younger patients? Geriatr Gerontol Int. 2021;21(1):60-5.
55. Zhou F, Yu T, Du R, Fan G, Liu Y, Liu Z, et al. Clinical course and risk factors for mortality of adult inpatients with COVID-19 in Wuhan, China: a retrospective cohort study. Lancet. 2020;395:1054-62.
56. Bravo F, Galvan G, Arancibia JM. Bacterial infections in COVID-19 patients hospitalized in intensive care unit. Rev Chil Infect. 2022;39(2):224-6.
57. Seidel J, Haller S, Eckmanns T, Harder T. Routine screening for colonization by Gram-negative bacteria in neonates at intensive care units for the prediction of sepsis: systematic review and meta-analysis. J Hosp Infect. 2018;99(4):367-80.
58. Aurilio C, Sansone P, Paladini A, Barbarisi M, Coppolino F, Pota V, et al. Multidrug resistence prevalence in COVID area. Life Basel Switz. 2021;11(7):1-8.
59. Huang C, Wang Y, Li X, Ren L, Zhao J, Hu Y, et al. Clinical features of patients infected with 2019 novel coronavirus in Wuhan, China. Lancet. 2020;395:497-506.
60. Chen N, Zhou M, Dong X, Qu J, Gong F, Han Y, et al. Epidemiological and clinical characteristics of 99 cases of 2019 novel coronavirus pneumonia in Wuhan, China: a descriptive study. Lancet. 2020;395:507-13.
61. Lai CC, Wang CY, Hsueh PR. Co-infections among patients with COVID- 19: The need for combination therapy with non-anti-SARS-CoV-2 agents? J Microbiol Immunol Infect. 2020;53(4):505-12.
62. Ding Q, Lu P, Fan Y, Xia Y, Liu M. The clinical characteristics of pneumonia patients coinfected with 2019 novel coronavirus and influenza virus in Wuhan, China. J Med Virol. 2020;92(9):1549-55.
63. Li H, Chen K, Liu M, Xu H, Xu Q. The profile of peripheral blood

lymphocyte subsets and serum cytokines in children with 2019 novel coronavirus pneumonia. J Infect. 2020;81(1):115-20.
64. Kim D, Quinn J, Pinsky B, Shah NH, Brown I. Rates of Co-infection Between SARS-CoV-2 and Other Respiratory Pathogens. JAMA. 2020;323(20):2085-6.
65. Rouze A, Martin-Loeches I, Povoa P, Makris D, Artigas A, Bouchereau M, et al. Relationship between SARS-CoV-2 infection and the incidence of ventilator-associated lower respiratory tract infections: a European multicenter cohort study. Intensive Care Med. 2021;47(2):188-98.
66. Husain M, Valayer S, Poey N, Rondinaud E, D'Humieres C, Visseaux B, et al. Co-infection and bacterial superinfection in patients hospitalised for COVID-19: a single-centre retrospective study. Med Mal Infect. 2020;50(6):S21.
67. Grasselli G, Cattaneo E, Florio G. Secondary infections in critically ill patients with COVID-19. Crit Care. 2021;25(1):1-6.
68. De Santis V, Corona A, Vitale D, Nencini C, Potalivo A, Prete A, et al. Bacterial infections in critically ill patients with SARS-2-COVID-19 infection: results of a prospective observational multicenter study. Infection. 2022;50(1):139-48.
69. Jarding EK, Flynn Makic MB. Central line care and management: adopting evidence-based nursing interventions. J Perianesthesia Nurs. 2021;36(4):328-33.
70. Costa RL, Lamas C, Simvoulidis LF, Espanha CA, Moreira LP, Bonancim RA, et al. Secondary infections in a cohort of patients with COVID-19 admitted to an intensive care unit: impact of gram-negative bacterial resistance. Rev Inst Med Trop Sao Paulo. 2022;64:1-10.
71. Lucien MA, Canarie MF, Kilgore PE, Jean-Denis G, Fenelon N, Pierre M, et al. Antibiotics and antimicrobial resistance in the COVID-19 era: Perspective from resource-limited settings. Int J Infect Dis. 2021;104:250-4.
72. Perez S, Innes GK, Walters MS, Mehr J, Arias J, Greeley R, et al. Increase in hospital-acquired carbapenem-resistant acinetobacter baumannii infection and colonization in an acute care hospital during a surge in COVID-19 admissions. Morb Mortal Wkly Rep. 2020;69(48):1827-31.
73. Arbune M, Decusara M, Macovei LA, Romila A, lancu AV, Indrei LL, et al. Surveillance of antibiotic resistance among Enterobacteriaceae strains isolated in an infectious diseases hospital from Romania. Rev Chim. 2018;69:1240-3.
74. Ebongue CO, Tsiazok MD, Mefo'o JP, Ngaba GP, Beyiha G, Adiogo D.

Evolution of antibiotic resistance of Enterobacteriaceae isolated at the Douala general Hospital from 2005 to 2012. Pan Afr Med J. 2015;20:1-11.

75. Michel-Briand Y. Resistance to the latest beta-lactams: mechanisms of acquisition and spread of resistance in Enterobacteriacea]. Bull Acad Natl Med. 2007;191(1):35-50;

76. Ruppe E, Woerther PL, Barbier F. Mechanisms of antimicrobial resistance in Gram-negative bacilli. Ann Intensive Care. 2015;5(1):1-15.

77. Shah PM, Stille W. Escherichia coli and Klebsiella pneumoniae strains more susceptible to cefoxitin than to third generation cephalosporins. J Antimicrob Chemother. 1983;11(6):597-8.

78. Ben Redjeb S, Ben Yaghlane H, Boujnah A, Philippon A, Labia R. Synergy between clavulanic acid and newer beta-lactams on nine clinical isolates of Klebsiella pneumoniae, Escherichia coli and Salmonella typhimurium resistant to third-generation cephalosporins. J Antimicrob Chemother. 1988;21(2):263-6.

79. Hammami A, Arlet G, Ben Redjeb S, Grimont F, Ben Hassen A, Rekik A, et al. Nosocomial outbreak of acute gastroenteritis in a neonatal intensive care unit in Tunisia caused by multiply drug resistant Salmonella wien producing SHV-2 beta-lactamase. Eur J Clin Microbiol Infect Dis. 1991;10(8):641-6.

80. Rhimi-Mahjoubi F, Bernier M, Arlet G, Jemaa ZB, Jouve P, Hammami A, et al. Identification of plasmid-encoded cephalosporinase ACC-1 among various enterobacteria (Klebsiella pneumoniae, Proteus mirabilis, Salmonella) isolated from a Tunisian hospital (Sfax 997-2000). Pathol Biol. 2002;50(1):7-11.

81. Fluit AC, Jones ME, Schmitz FJ, Acar J, Gupta R, Verhoef J. Antimicrobial susceptibility and frequency of occurrence of clinical blood isolates in Europe from the SENTRY antimicrobial surveillance program, 1997 and 1998. Clin Infect Dis. 2000;30(3):454-60.

82. Diekema DJ, Pfaller MA, Jones RN, Doern GV, Winokur PL, Gales AC, et al. Survey of bloodstream infections due to gram-negative bacilli: frequency of occurrence and antimicrobial susceptibility of isolates collected in the United States, Canada, and Latin America for the sentry antimicrobial surveillance program, 1997. Clin Infect Dis. 1999;29(3):595-607.

83. Almeida TL, Mendo T, Costa R, Novais C, Margal M, Martins F, et al. Carbapenemase-producing enterobacteriaceae (CPE) newborn colonization in a portuguese neonatal intensive care unit (NICU): Epidemiology and infection prevention and control measures. Infect Dis Rep. 2021;13(2):411- 7.

84. Beigverdi R, Sattari-Maraji A, Jabalameli F, Emaneini M. Prevalence of genes encoding aminoglycoside-modifying enzymes in clinical isolates of gram-positive cocci in Iran: A systematic review and meta-analysis. Microb Drug Resist. 2020;26(2):126-35.
85. Alvarez-Lerma F, Grau S, Gracia-Arnillas MP. Gram-positive cocci infections in intensive care: guide to antibacterial selection. Drugs. 2006;66(6):751-68.
86. Luo XQ, Gong YL, Zhang C, Liu MX, Shi YL, Peng YZ, et al. Analysis of distribution and drug resistance of pathogens isolated from 159 patients with catheter-related bloodstream infection in burn intensive care unit. Zhonghua Shao Shang Za Zhi. 2020;36(1):24-31.
87. Ebongue CO, Mengue ER, Mefo'o J, Tsiazok MD, Kouassi RN, Bum EN. Antimicrobial multi-resistance of acinetobacter baumanii isolated from clinical specimens in Douala (Cameroon). J Dis Med Plants. 2015;1(2):31- 6.
88. Gitau W, Masika M, Musyoki M, Museve B, Mutwiri T. Antimicrobial susceptibility pattern of Staphylococcus aureus isolates from clinical specimens at Kenyatta national Hospital. BMC Res Notes. 2018;11(1):1-5.
89. Aissou M, Bousnane NEH, Kaoueche S. Etude de la résistance aux antibiotiques des souches Staphylococcus aureus isolees du milieu hospitalier [En Ligne]. 2018 [Accessed 05/07/2023]; Available from: http://localhost:8080/xmlui/handle/123456789/5447
90. Azzab MM, El-Sokkary RH, Tawfeek MM, Gebriel MG. Multidrugresistant bacteria among patients with ventilatorassociated pneumonia in an emergency intensive care unit, Egypt. East Mediterr Health. 2017;22(12):894-903.
91. Krir A, Dhraief S, Messadi AA, Thabet L. Bacteriological profile and antibiotic resistance of bacteria isolated in a burns resuscitation unit over seven years. Ann Burns Fire Disasters. 2019;32(3):197-202.
92. Deyno S, Toma A, Worku M, Bekele M. Antimicrobial resistance profile of staphylococcus aureus isolates isolated from ear discharges of patients at University of Hawassa comprehensive specialized hospital. BMC Pharmacol Toxicol. 2017;18(1):1-7.
93. Tran GM, Ho-Le TP, Ha DT, Tran-Nguyen CH, Nguyen TS, Pham TT, et al. Patterns of antimicrobial resistance in intensive care unit patients: a study in Vietnam. BMC Infect Dis. 2017;17(1):1-7.
94. Moutaouakkil Y, Siah S, Bennana A, Tadlaoui Y, Makram S, Cherrah Y, et al. Reevaluation clinico-biologique de I'antibiotherapie probabiliste en reanimation des brules. Ann Burns Fire Disasters. 2018;31(1):35-41.

95. Shrestha P, Cooper BS, Coast J, Oppong R, Do Thi Thuy N, Phodha T, et al. Enumerating the economic cost of antimicrobial resistance per antibiotic consumed to inform the evaluation of interventions affecting their use. Antimicrob Resist Infect Control. 2018;7:1-9.
96. Wu C, Chen X, Cai Y, Xia J, Zhou X, Xu S, et al. Risk factors associated with acute respiratory distress syndrome and death in patients with coronavirus disease 2019 pneumonia in Wuhan, China. JAMA Intern Med. 2020;180(7):934-43.
97. Grasselli G, Zangrillo A, Zanella A, Antonelli M, Cabrini L, Castelli A, et al. Baseline characteristics and outcomes of 1591 patients infected with SARS-CoV-2 admitted to ICUs of the lombardy region, Italy. JAMA. 2020;323(16):1574-81.
98. Lansbury L, Lim B, Baskaran V, Lim WS. Co-infections in people with COVID-19: a systematic review and meta-analysis. J Infect. 2020;81(2):266-75.
99. MacVane SH. Antimicrobial Resistance in the intensive care unit: A focus on gram-negative bacterial infections. J Intensive Care Med. 2017;32(1):25- 37.
100. Klein EY, Monteforte B, Gupta A, Jiang W, May L, Hsieh YH, et al. The frequency of influenza and bacterial coinfection: a systematic review and meta-analysis. Influenza Other Respir Viruses. 2016;10(5):394-403.
101. Ayari H, Azouzi F, Hammouda A, Boughattas S, Tilouche L, Ketata S, et al. Microbiological profile of bacteremia in intensive care. Infect Dis Now. 2021;51:41-2.
102. Singhal T, Shah S, Thakkar P, Naik R. The incidence, aetiology and antimicrobial susceptibility of central line-associated bloodstream infections in intensive care unit patients at a private tertiary care hospital in Mumbai, India. Indian J Med Microbiol. 2019;37(4):521-6.
103. Djordjevic ZM, Folic MM, Jankovic SM. Distribution and antibiotic susceptibility of pathogens isolated from adults with hospital-acquired and ventilator-associated pneumonia in intensive care unit. J Infect Public Health. 2017;10(6):740-4.
104. Jamoussi A, Ayed S, Ben Ismail K, Chtara K, Bouaziz M, Mokline A, et al. The prevalence of healthcare-associated infection in medical intensive care units in Tunisia. Results of the multi-centre nosorea1 study. Tunis Med. 2018;96:731-6.
105. Solmaz i, Kalin BS. Assessment of antibiotic resistance of infectious agents in patients with pneumonia in tertiary critical care unit and effect on

clinical outcomes. Int J Clin Pract. 2021;75(4):1-17.
106. Kumari P, Datta P, Gombar S, Sharma D, Chander J. Epidemiology and clinical outcome of ventilator-associated events at a tertiary care hospital from North India. Trop Doct. 2021;51(2):162-7.
107. Bailee YG, Acer O, Ozudogru O. Evaluation of bacterial agents isolated from endotracheal aspirate cultures of COVID-19 general intensive care patients and their antibiotic resistance profiles compared to pre-pandemic conditions. Microb Pathog. 2022;164:1-7.

MIX
Papier aus verantwortungsvollen Quellen
Paper from responsible sources
FSC® C105338

Printed by Books on Demand GmbH, Norderstedt / Germany